AGRICULTURE METEOROLOGY

AGRICULTURE METEOROLOGY

Dr. Parmeshwar Singh

RANDOM

RANDOM PUBLICATIONS

NEW DELHI - 110 002 (INDIA)

Agriculture Meteorology

ISBN 978-93-51117-37-7

Published in 2015 in India by

RANDOM PUBLICATIONS

4376-A/4B, Gali Murari Lal, Ansari Road
New Delhi-110 002
Phone: +9111-43580356, 23289044
E-mail: randomexports@gmail.com; sales@randompublications.com; info@randompublications.com

Reprinted 2026

Type Setting by: Friends Media, Delhi-110089
Printed at : Replika Press Pvt. Ltd.

Preface

Meteorology is the interdisciplinary scientific study of the atmosphere. Studies in the field stretch back millennia, though significant progress in meteorology did not occur until the 18th century. The 19th century saw modest progress in the field after observing networks formed across several countries. It wasn't until after the development of the computer in the latter half of the 20th century that significant breakthroughs in weather forecasting were achieved. Meteorological phenomena are observable weather events which illuminate, and are explained by the science of meteorology. Those events are bound by the variables that exist in Earth's atmosphere; temperature, air pressure, water vapour, and the gradients and interactions of each variable, and how they change in time. Different spatial scales are studied to determine how systems on local, regional, and global levels impact weather and climatology.

Meteorology, climatology, atmospheric physics, and atmospheric chemistry are sub-disciplines of the atmospheric sciences. Meteorology and hydrology compose the interdisciplinary field of hydrometeorology. Interactions between Earth's atmosphere and the oceans are part of coupled ocean-atmosphere studies. Meteorology has application in many diverse fields such as the military, energy production, transport, agriculture and construction. Each science has its own unique sets of laboratory equipment. In the atmosphere, there are many things or qualities of the atmosphere that can be measured. Rain, which can be observed, or seen anywhere and anytime was one of the first ones to be measured historically. Also, two other accurately measured qualities are wind and humidity. Neither of these can be seen but can be felt. The devices to measure these three sprang up in the mid-15th century and were respectively the rain gauge, the anemometer, and the hygrometer. Many attempts had been made prior to the 15th century to construct adequate equipment to measure the many atmospheric variables. Many were faulty in some way or were simply not reliable. Even Aristotle noted this in some of his work; as the difficulty to measure the air. Sets of surface measurements are important data to meteorologists. They give a snapshot of a variety of weather conditions at one single location and are usually at a weather station, a ship or a weather buoy. The measurements taken at a weather station can include any number of atmospheric observables. Usually, temperature, pressure, wind measurements, and humidity are the variables that are measured by a thermometer, barometer, anemometer, and hygrometer, respectively. Upper air data are of crucial importance for weather forecasting. The most widely used technique is launches of radiosondes. Supplementing the radiosondes a network of aircraft collection is organized by the World Meteorological Organization. Remote sensing, as used in meteorology, is the concept of collecting data from remote weather events and subsequently producing weather information. The common types of remote sensing are Radar, Lidar, and satellites (or photogrammetry). Each collects data about the atmosphere from a remote location and, usually, stores the data where the instrument is located. Radar and Lidar are not passive because both use EM radiation to illuminate a specific portion of the atmosphere. Weather satellites along with more general-purpose Earth-observing satellites circling the earth at various altitudes have become an indispensable tool for studying a wide range of phenomena from forest fires to El Niño. In the study of the atmosphere, meteorology can be divided into distinct areas of emphasis depending on the temporal scope and spatial scope of interest. At one extreme of this scale is climatology. In the timescales of hours to days, meteorology separates into micro-, meso-, and synoptic scale meteorology. Respectively, the geospatial size of each of these three scales relates directly with the appropriate timescale. Other subclassifications are available based on the need by or by the unique, local or broad effects that are studied within that sub-class. Weather forecasting is the application of science and technology to predict the state of the atmosphere for a future time and a given location.

The aim of this book is to provide, in a convenient form, information regarding the most important practices and procedures in agricultural meteorology. This succinct, compact introduction to meteorology is designed especially for those with minimal background in science and mathematics. This comprehensive book will satisfy all who need to know more about the weather and climate of arid lands.

I thank all members of my team who have helped in the preparation of the book. My special thanks go to "Random Publications" who have published the book.

— Dr. Parmeshwar Singh

Contents

Chapter 1

Introduction

Meteorology

Meteorology is the interdisciplinary scientific study of the atmosphere. Studies in the field stretch back millennia, though significant progress in meteorology did not occur until the 18th century. The 19th century saw modest progress in the field after observing networks formed across several countries. It wasn't until after the development of the computer in the latter half of the 20th century that significant breakthroughs in weather forecasting were achieved.

Meteorological phenomena are observable weather events which illuminate, and are explained by the science of meteorology. Those events are bound by the variables that exist in Earth's atmosphere; temperature, air pressure, water vapour, and the gradients and interactions of each variable, and how they change in time. Different spatial scales are studied to determine how systems on local, regional, and global levels impact weather and climatology.

Meteorology, climatology, atmospheric physics, and atmospheric chemistry are sub-disciplines of the atmospheric sciences. Meteorology and hydrology compose the interdisciplinary field of hydrometeorology. Interactions between Earth's atmosphere and the oceans are part of coupled ocean-atmosphere studies. Meteorology has application in many diverse fields such as the military, energy production, transport, agriculture and construction.

History

The beginnings of meteorology can be traced back to ancient India, as the Upanishads contain serious discussion about the processes of cloud formation and rain and the seasonal cycles caused by the movement of earth around the sun. Varâhamihira's classical work *Brihatsamhita*, written about 500 AD, provides clear evidence that a deep knowledge of atmospheric processes existed even in those times.

In 350 BC, Aristotle wrote *Meteorology*. Aristotle is considered the founder of meteorology. One of the most impressive achievements described in the *Meteorology* is the description of what is now known as the hydrologic cycle. The Greek scientist Theophrastus compiled a book on weather forecasting, called the *Book of Signs*.

The work of Theophrastus remained a dominant influence in the study of weather and in weather forecasting for nearly 2,000 years. In 25 AD, Pomponius Mela, a geographer for the Roman Empire, formalized the climatic zone system. According to Toufic Fahd, around the 9th century, Al-Dinawari wrote the *Kitab al-Nabat* (*Book of Plants*), in which he deals with the application of meteorology to agriculture during the Muslim Agricultural Revolution. He describes the meteorological character of the sky, the planets and constellations, the sun and moon, the lunar phases indicating seasons and rain, the *anwa* (heavenly bodies of rain), and atmospheric phenomena such as winds, thunder, lightning, snow, floods, valleys, rivers, lakes.

Research of Visual Atmospheric Phenomena

Ptolemy wrote on the atmospheric refraction of light in the context of astronomical observations. In 1021, Alhazen showed that atmospheric refraction is also responsible for twilight; he estimated that twilight begins when the sun is 19 degrees below the horizon, and also used a geometric determination based on this to estimate the maximum possible height of the earth's atmosphere as 52,000 *passuum* (about 49 miles, or 79 km).

St. Albert the Great was the first to propose that each drop of falling rain had the form of a small sphere, and that this form meant that the rainbow was produced by light interacting with each raindrop. Roger Bacon was the first to calculate the angular size of the rainbow. He stated that the rainbow summit can not appear higher than 42 degrees above the horizon.

In the late 13th century and early 14th century, Kamâl al-Dîn al-Fârisî and Theodoric of Freiberg were the first to give the correct explanations for the primary rainbow phenomenon. Theoderic went further and also explained the secondary rainbow. In 1716, Edmund Halley suggested that aurorae are caused by "magnetic effluvia" moving along the Earth's magnetic field lines.

Instruments and Classification Scales

In 1441, King Sejong's son, Prince Munjong, invented the first standardized rain gauge. These were sent throughout the Joseon Dynasty of Korea as an official tool to assess land taxes based upon a farmer's potential harvest. In 1450, Leone Battista Alberti developed a swinging-plate anemometre, and was known as the first *anemometre.* In 1607, Galileo Galilei constructed a thermoscope. In 1611, Johannes Kepler wrote the first scientific treatise on snow crystals: "Strena Seu de Nive Sexangula (A New Year's Gift of Hexagonal Snow)". In 1643, Evangelista Torricelli invented the mercury barometre.

In 1662, Sir Christopher Wren invented the mechanical, self-emptying, tipping bucket rain gauge. In 1714, Gabriel Fahrenheit created a reliable scale for measuring temperature with a mercury-type thermometre. In 1742, Anders Celsius, a Swedish astronomer, proposed the "centigrade" temperature scale, the predecessor of the current Celsius scale. In 1783, the first hair hygrometre was demonstrated by Horace-Bénédict de Saussure. In 1802–1803, Luke Howard wrote *On the Modification of Clouds* in which he assigns cloud types Latin names. In 1806, Francis Beaufort introduced his system for classifying wind speeds. Near the end of the 19th century the first cloud atlases were published, including the *International Cloud Atlas*, which has remained in print ever since. The April 1960 launch of the first successful weather satellite, TIROS-1, marked the beginning of the age where weather information became available globally.

Atmospheric Composition Research

In 1648, Blaise Pascal rediscovered that atmospheric pressure decreases with height, and deduced that there is a vacuum above the atmosphere. In 1738, Daniel Bernoulli published *Hydrodynamics*, initiating the kinetic theory of gases and established the basic laws for the theory of gases. In 1761, Joseph Black discovered that ice absorbs heat without changing its temperature when melting. In 1772, Black's student Daniel Rutherford discovered nitrogen, which he called *phlogisticated air*, and together they developed the phlogiston theory. In 1777, Antoine Lavoisier discovered oxygen and developed an explanation for combustion. In 1783, in Lavoisier's book *Reflexions sur le phlogistique*, he deprecates the phlogiston theory and proposes a caloric theory. In 1804, Sir John Leslie observed that a matte black surface radiates heat more effectively than a polished surface, suggesting the importance of black body radiation. In 1808, John Dalton defended caloric theory in *A New System of Chemistry* and described how it combines with matter, especially gases; he proposed that the heat capacity of gases varies inversely with atomic weight. In 1824, Sadi Carnot analyzed the efficiency of steam engines using caloric theory; he developed the notion of a reversible process and, in postulating that no such thing exists in nature, laid the foundation for the second law of thermodynamics.

Research into Cyclones and Air Flow

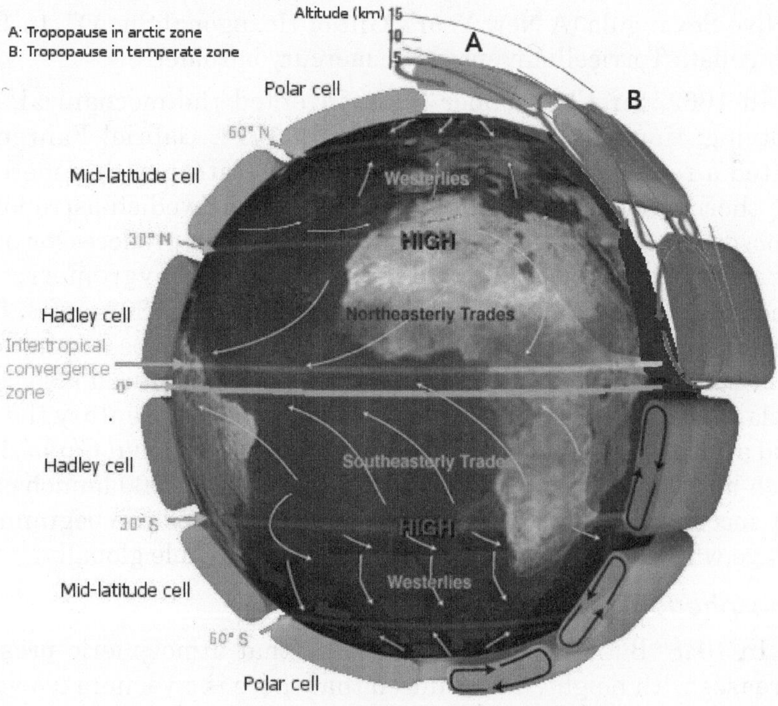

***Figure:** The westerlies and trade winds are part of the earth's atmospheric circulation*

In 1494, Christopher Columbus experienced a tropical cyclone, which led to the first written European account of a hurricane. In 1686, Edmund Halley presented a systematic study of the trade winds and monsoons and identified solar heating as the cause of atmospheric motions. In 1735, an *ideal* explanation of global circulation through study of the trade winds was written by George Hadley. In 1743, when Benjamin Franklin was prevented from seeing a lunar eclipse by a hurricane, he decided that cyclones move in a contrary manner to the winds at their periphery. Understanding the kinematics of how exactly the rotation of the earth affects airflow was partial at first. Gaspard-Gustave Coriolis published a paper in 1835 on the energy yield of machines with rotating parts, such as waterwheels. In 1856, William Ferrel proposed the existence of a circulation cell in the mid-latitudes, with air being deflected by the Coriolis force to create the prevailing westerly winds.

Late in the 19th century, the full extent of the large-scale interaction of pressure gradient force and deflecting force that in the end causes air masses to move *along* isobars was understood. By 1912, this deflecting force was named the Coriolis effect. Just after World War I, a group of meteorologists in Norway led by Vilhelm Bjerknes developed the Norwegian cyclone model that explains the generation, intensification and ultimate decay (the life cycle) of mid-latitude cyclones, introducing the idea of fronts, that is, sharply defined boundaries between air masses. The group included Carl-Gustaf Rossby (who was the first to explain the large scale atmospheric flow in terms of fluid dynamics), Tor Bergeron (who first determined the mechanism by which rain forms) and Jacob Bjerknes.

Observation Networks and Weather Forecasting

In 1654, Ferdinando II de Medici established the first *weather observing* network, that consisted of meteorological stations in Florence, Cutigliano, Vallombrosa, Bologna, Parma, Milan, Innsbruck, Osnabrück, Paris and Warsaw. Collected data were centrally sent to Florence at regular time intervals. In 1832, an electromagnetic telegraph was created by Baron Schilling. The arrival of the electrical telegraph in 1837 afforded, for the first time, a practical method for quickly gathering surface weather observations from a wide area. This data could be used to produce maps of the state of the atmosphere for a region near the earth's surface and to study how these states evolved through time.

To make frequent weather forecasts based on these data required a reliable network of observations, but it was not until 1849 that the Smithsonian Institution began to establish an observation network across the United States under the leadership of Joseph Henry. Similar observation networks were established in Europe at this time. In 1854, the United Kingdom government appointed Robert FitzRoy to the new office of *Meteorological Statist to the Board of Trade* with the role of gathering weather observations at sea. FitzRoy's office became the United Kingdom Meteorological Office in 1854, the first national meteorological service in the world. The first daily weather forecasts made by FitzRoy's Office were published in *The Times* newspaper in 1860. The following year a system was introduced of hoisting storm warning cones at principal ports when a gale was expected.

Over the next 50 years many countries established national meteorological services. The India Meteorological Department (1875) was established following tropical cyclone and monsoon related famines in the previous decades. The Finnish Meteorological Central Office (1881) was formed from part of Magnetic Observatory of Helsinki University. Japan's Tokyo Meteorological Observatory, the forerunner of the Japan Meteorological Agency, began constructing surface weather maps in 1883. The United States Weather Bureau (1890) was established under the United States Department of Agriculture. The Australian Bureau of Meteorology (1906) was established by a Meteorology Act to unify existing state meteorological services.

Numerical Weather Prediction

In 1904, Norwegian scientist Vilhelm Bjerknes first argued in his paper *Weather Forecasting as a Problem in Mechanics and Physics* that it should be possible to forecast weather from calculations based upon natural laws.

It was not until later in the 20th century that advances in the understanding of atmospheric physics led to the foundation of modern numerical weather prediction. In 1922, Lewis Fry Richardson published "Weather Prediction By Numerical Process", after finding notes and derivations he worked on as an ambulance driver in World War I. He described therein how small terms in the prognostic fluid dynamics equations governing atmospheric flow could be neglected, and a finite differencing scheme in time and space could be devised, to allow numerical prediction solutions to be found. Richardson envisioned a large auditorium of thousands of people performing the calculations and passing them to others. However, the sheer number of calculations required was too large to be completed without the use of computers, and the size of the grid and time steps led to unrealistic results in deepening systems. It was later found, through numerical analysis, that this was due to numerical instability.

Starting in the 1950s, numerical forecasts with computers became feasible. The first weather forecasts derived this way used barotropic (single-vertical-level) models, and could successfully predict the large-scale movement of midlatitude Rossby waves, that is, the pattern of atmospheric lows and highs.

In the 1960s, the chaotic nature of the atmosphere was first observed and mathematically described by Edward Lorenz, founding the field of chaos theory. These advances have led to the current use of ensemble forecasting in most major forecasting centres, to take into account uncertainty arising from the chaotic nature of the atmosphere. Climate models have been developed that feature a resolution comparable to older weather prediction models. These climate models are used to investigate long-term climate shifts, such as what effects might be caused by human emission of greenhouse gases.

Meteorologists

Meteorologists are scientists who study meteorology. The American Meteorological Society published and continually updates an authoritative electronic *Meteorology Glossary*. Meteorologists work in government agencies, private consulting and research services, industrial enterprises, utilities, radio and television stations, and in education. In the United States, meteorologists held about 9,400 jobs in 2009.

Meteorologists are best known for forecasting the weather. Many radio and television weather forecasters are professional meteorologists, while others are reporters (weather specialist, weatherman, etc.) with no formal meteorological training. The American Meteorological Society and National Weather Association issue "Seals of Approval" to weather broadcasters who meet certain requirements.

Equipment

Each science has its own unique sets of laboratory equipment. In the atmosphere, there are many things or qualities of the atmosphere that can be measured. Rain, which can be observed, or seen anywhere and anytime was one of the first ones to be measured historically. Also, two other accurately measured qualities are wind and humidity. Neither of these can be seen but can be felt. The devices to measure these three sprang up in the mid-15th century and were respectively the rain gauge, the anemometre, and the hygrometre. Many attempts had been made prior to the 15th century to construct adequate equipment to measure the many atmospheric variables. Many were faulty in some way or were simply not reliable. Even Aristotle noted this in some of his work; as the difficulty to measure the air.

Sets of surface measurements are important data to meteorologists. They give a snapshot of a variety of weather conditions at one single location and are usually at a weather station, a ship or a weather buoy. The measurements taken at a weather station can include any number of atmospheric observables. Usually, temperature, pressure, wind measurements, and humidity are the variables that are measured by a thermometre, barometre, anemometre, and hygrometre, respectively. Upper air data are of crucial importance for weather forecasting. The most widely used technique is launches of radiosondes. Supplementing the radiosondes a network of aircraft collection is organized by the World Meteorological Organization.

Remote sensing, as used in meteorology, is the concept of collecting data from remote weather events and subsequently producing weather information. The common types of remote sensing are Radar, Lidar, and satellites (or photogrammetry). Each collects data about the atmosphere from a remote location and, usually, stores the data where the instrument is located. Radar and Lidar are not passive because both use EM radiation to illuminate a specific portion of the atmosphere. Weather satellites along with more general-purpose Earth-observing satellites circling the earth at various altitudes have become an indispensable tool for studying a wide range of phenomena from forest fires to El Niño.

Spatial Scales

In the study of the atmosphere, meteorology can be divided into distinct areas of emphasis depending on the temporal scope and spatial scope of interest. At one extreme of this scale is climatology. In the timescales of hours to days, meteorology separates into micro-, meso-, and synoptic scale meteorology. Respectively, the geospatial size of each of these three scales relates directly with the appropriate timescale.

Other subclassifications are available based on the need by or by the unique, local or broad effects that are studied within that sub-class.

Microscale

Microscale meteorology is the study of atmospheric phenomena of about 1 km or less. Individual thunderstorms, clouds, and local turbulence caused by buildings and other obstacles, such as individual hills fall within this category.

Mesoscale

Mesoscale meteorology is the study of atmospheric phenomena that has horizontal scales ranging from microscale limits to synoptic scale limits and a vertical scale that starts at the Earth's surface and includes the atmospheric boundary layer, troposphere, tropopause, and the lower section of the stratosphere. Mesoscale timescales last from less than a day to the lifetime of the event, which in some cases can be weeks. The events typically of interest are thunderstorms, squall lines, fronts, precipitation bands in tropical and extratropical cyclones, and topographically generated weather systems such as mountain waves and sea and land breezes.

Synoptic Scale

Synoptic scale meteorology is generally large area dynamics referred to in horizontal coordinates and with respect to time. The phenomena typically described by synoptic meteorology include events like extratropical cyclones, baroclinic troughs and ridges, frontal zones, and to some extent jet streams. All of these are typically given on weather maps for a specific time. The minimum horizontal scale of synoptic phenomena is limited to the spacing between surface observation stations.

Global scale

Global scale meteorology is study of weather patterns related to the transport of heat from the tropics to the poles. Also, very large scale oscillations are of importance. These oscillations have time periods typically on the order of months, such as the Madden-Julian Oscillation, or years, such as the El Niño-Southern Oscillation and the Pacific decadal oscillation. Global scale pushes the thresholds of the perception of meteorology into climatology. The traditional definition of climate is pushed into larger timescales with the further understanding of how the global oscillations cause both climate and weather disturbances in the synoptic and mesoscale timescales.

Numerical Weather Prediction is a main focus in understanding air–sea interaction, tropical meteorology, atmospheric predictability, and tropospheric/stratospheric processes. The Naval Research Laboratory in Monterey California developed a global atmospheric model called Navy Operational Global Atmospheric Prediction System (NOGAPS). NOGAPS is run operationally at Fleet Numerical Meteorology and Oceanography Centre for the United States Military. Many other global atmospheric models are run by national meteorological agencies.

Some Meteorological Principles

Boundary Layer Meteorology

Boundary layer meteorology is the study of processes in the air layer directly above earth's surface, known as the atmospheric boundary layer (ABL). The effects of the surface – heating, cooling, and friction – cause turbulent mixing within the air layer. Significant fluxes of heat, matter, or momentum on time scales of less than a day are advected by turbulent motions. Boundary layer meteorology includes the study of all types of surface–atmosphere boundary, including ocean, lake, urban land and non-urban land for the study of meteorology.

Dynamic Meteorology

Dynamic meteorology generally focuses on the fluid dynamics of the atmosphere. The idea of air parcel is used to define the smallest element of the atmosphere, while ignoring the discrete molecular and chemical nature of the atmosphere. An air parcel is defined as a point in the fluid continuum of the atmosphere. The fundamental laws of fluid dynamics, thermodynamics, and motion are used to study the atmosphere. The physical quantities that characterize the state of the atmosphere are temperature, density, pressure, etc. These variables have unique values in the continuum.

Applications

Weather Forecasting

Weather forecasting is the application of science and technology to predict the state of the atmosphere for a future time and a given location. Human beings have attempted to predict the weather informally for millennia, and formally since at least the 19th century. Weather forecasts are made by collecting quantitative data about the current state of the atmosphere and using scientific understanding of atmospheric processes to project how the atmosphere will evolve.

Once an all-human endeavour based mainly upon changes in barometric pressure, current weather conditions, and sky condition, forecast models are now used to determine future conditions. Human input is still required to pick the best possible forecast model to base the forecast upon, which involves pattern recognition skills, teleconnections, knowledge of model performance, and knowledge of model biases. The chaotic nature of the atmosphere, the massive computational power required to solve the equations that describe the atmosphere, error involved in measuring the initial conditions, and an incomplete understanding of atmospheric processes mean that forecasts become less accurate as the difference in current time and the time for which the forecast is being made (the *range* of the forecast) increases. The use of ensembles and model consensus help narrow the error and pick the most likely outcome.

There are a variety of end uses to weather forecasts. Weather warnings are important forecasts because they are used to protect life and property. Forecasts based on temperature and precipitation are important to agriculture, and therefore to commodity traders within stock markets. Temperature forecasts are used by utility companies to estimate demand over coming days. On an everyday basis, people use weather forecasts to determine what to wear on a given day. Since outdoor activities are severely curtailed by heavy rain, snow and the wind chill, forecasts can be used to plan activities around these events, and to plan ahead and survive them.

Aviation Meteorology

Aviation meteorology deals with the impact of weather on air traffic management. It is important for air crews to understand the implications of weather on their flight plan as well as their aircraft, as noted by the Aeronautical Information Manual:

> *The effects of ice on aircraft are cumulative—thrust is reduced, drag increases, lift lessens, and weight increases. The results are an increase in stall speed and a deterioration of aircraft performance. In extreme cases, 2 to 3 inches of ice can form on the leading edge of the airfoil in less than 5 minutes. It takes but 1/2 inch of ice to reduce the lifting power of some aircraft by 50 percent and increases the frictional drag by an equal percentage.*

Agricultural Meteorology

Meteorologists, soil scientists, agricultural hydrologists, and agronomists are persons concerned with studying the effects of weather and climate on plant distribution, crop yield, water-use efficiency, phenology of plant and animal development, and the energy balance of managed and natural ecosystems. Conversely, they are interested in the role of vegetation on climate and weather.

Hydrometeorology

Hydrometeorology is the branch of meteorology that deals with the hydrologic cycle, the water budget, and the rainfall statistics of storms. A hydrometeorologist prepares and issues forecasts of accumulating (quantitative) precipitation, heavy rain, heavy snow, and highlights areas with the potential for flash flooding. Typically the range of knowledge that is required overlaps with climatology, mesoscale and synoptic meteorology, and other geosciences.

The multidisciplinary nature of the branch can result in technical challenges, since tools and solutions from each of the individual disciplines involved may behave slightly differently, be optimized for different hard- and software platforms and use different data formats. There are some initiatives - such as the DRIHM project - that are trying to address this issue.

Nuclear Meteorology

Nuclear meteorology investigates the distribution of radioactive aerosols and gases in the atmosphere.

Maritime Meteorology

Maritime meteorology deals with air and wave forecasts for ships operating at sea. Organizations such as the Ocean Prediction Centre, Honolulu National Weather Service forecast office, United Kingdom Met Office, and JMA prepare high seas forecasts for the world's oceans.

Military Meteorology

Military meteorology is the research and application of meteorology for military purposes. In the United States, the United States Navy's Commander, Naval Meteorology and Oceanography Command oversees meteorological efforts for the Navy and Marine Corps while the United States Air Force's Air Force Weather Agency is responsible for the Air Force and Army.

Agricultural Meteorology and its Levels

1. J.W. Smith (1916) has defined Agricultural Meteorology as "Meteorology in its relation to agriculture"
2. It can be defined as the science investigating Meteorology, climates and hydrologic condition, which are significant to agriculture.
3. In short Agril. Meteorology is the applied branch of meteorology, which deals with the relationship between climates, weather and life and growth of the cultivated plants and animals.

Levels of Study of Meteorology

Study of meteorology is organized at three levels.

Micro scale: A process operating within vegetation canopies near earth surface its size is in few cm and Life span is few seconds.

Mesoscale: The systems are approximately 10km in size and a lifetime is of few hours [up to5 hrs] eg. Thunder storm.

Macro scale: It is divided into two scales.

Synoptic scale: These systems have a diametre of few thousand km. and life time of about 5 days

Eg. Tropical storm, cyclones.

Planetary scale: These systems have a diametre of 5000 to 10000km and persist for several weeks

Scope of Meteorology

Meteorology is a branch of science which deals with the methodical study of the atmosphere and the effects it bears on the day-to-day activities of the inhabitants. The scope of this particular subject has increased to a great extent because of the alarming and dangerous enhancements in the pollution levels. The effects that the increased pollution has on us are leading to the far reaching ill effects and spread of a large number of ailments and diseases.

The aggravation in the environmental hazards such as global warming, forest cover reduction, erosion etc. has provided this subject the importance it didn't enjoy in the past. The study of meteorology comes under the study of earth sciences that is taught at a number of technical institutes and also in various universities of the country. Some of the prominent institutes which provide several courses in the field of meteorology in various levels of academics are Indian Institute of Science, Aryabhatta Research Institute of Observational Sciences, Indian Institute of Tropical Meteorology, and so on.

Importance of Meteorology

Most of the industrial, social, agricultural and commercial activities of the economy are directly or indirectly affected by the weather and climate. Also the atmosphere influences and upholds all living organisms including human life, micro- organisms, animals, pests, insects, plants and marine culture at all times. Meteorology has therefore the greatest scope on every single activity of life. Some of the fields of applications of the enhanced field of science are safe navigation, safe aviation, fisheries, land use planning, commerce and so on.

Importance and Scope of Meteorology in Agriculture

Weather and climate is a resource and considered as basic input or resources in agricultural planning, every plant process related with growth development and yield of a crop is affected by weather.

Similarly every farm operation such as ploughing harrowing, land preparation, weeding, irrigation, manuring, spraying, dusting, harvesting, threshing, storage and transport of farm produce are affected by weather. The scope of Agril Meteorology can be illustrated through the following few applications.

Characterization of Agricultural Climate

For determining crop growing season, solar radiation, air temperature, precipitation, wind, humidity etc. are important climatic factors on which the growth, development and yield of a crop depends Agro-meteorology considers and assess the suitability of these parametres in a given region for maximum crop production and economical benefits.

Crop Planning for Stability in Production

To reduce risk of crop failure on climatic part, so as to get stabilized yields even under weather adversity, suitable crops/cropping patterns/ contingent cropping planning can be selected by considering water requirements of crop, effective, rainfall and available soil moisture.

Crop Management

Management of crop involves various farm operations such as, sowing fertilizer application. Plat protection, irrigation scheduling, harvesting etc. can be carried out on the basis of specially tailored weather support. For this the use of operational forecasts, available from agro met advisories, is made e.g.:

1) Weeding harrowing, mulching etc. are undertaken during dry spells forecasted.
2) Fertilizer application is advisable when rainfall is not heavy wind speed is<30 km/hr and soil moisture is between 30 to 80%
3) Spraying/dusting is undertaken when there is no rainfall, soil moisture is 90% and wind speed is<25km/hr.

Crop Monitoring

To check crop health and growth performance of a crop, suitable meteorological tools such as crop growth models. Water balance technique or remote sensing etc. Can be used.

Crop Modelling and Yield–climate Relationship

Suitable crop models, devised for the purpose can provide information or predict the results about the growth and yield when the current and past weather data is used.

Research in Crop–climate Relationship

Agro-meteorology can help to understand crop-climate relationship so as to resolve complexities of plant process in relation to its micro climate.

Climate Extremities

Climatic extremities such a frost floods, droughts, hail storms, high winds can be forecasted and crop can be protected.

Climate as a Tool to Diagnose Soil Moisture Stress

Soil moisture can be exactly determined from climatic water balance method, Which is used to diagnose the soil moisture stress, drought and necessary protective measures such as irrigation, mulching application of antitranspirant, defoliation, thinning etc. can be undertaken.

Livestock Production

Livestock production is a part of agriculture. The set of favourable and unfavourable weather conditions for growth, development and production of livestock is livestock is studied in Agril. Meteorology. Thus to optimize milk production poultry production, the climatic normal are worked out and on the suitable breeds can be evolved or otherwise can provide the congenial conditions for the existing breeds.

Soil Formation

Soil formation process depend on climatic factors like temperature, precipitation, humidity, wind etc., thus climate is a major factor in soil formation and development.

Weather

Weather is the state of the atmosphere, to the degree that it is hot or cold, wet or dry, calm or stormy, clear or cloudy. Weather, seen from an anthropological perspective, is something all humans in the world constantly experience through their senses, at least while being outside. There are socially and scientifically constructed understandings of what weather is, what makes it change, what effects it has on humans in different situations etc. Therefore weather is something people often communicate about. Turning back to the meteorological perspective, most weather phenomena occur in the troposphere, just below the stratosphere. Weather generally refers to day-to-day temperature and precipitation activity, whereas climate is the term for the statistics of atmospheric conditions over longer periods of time. When used without qualification, "weather", is generally understood to mean the weather of Earth.

Weather is driven by air pressure (temperature and moisture) differences between one place and another. These pressure and temperature differences can occur due to the sun angle at any particular spot, which varies by latitude from the tropics. The strong temperature contrast between polar and tropical air gives rise to the jet stream. Weather systems in the mid-latitudes, such as extratropical cyclones, are caused by instabilities of the jet stream flow. Because the Earth's axis is tilted relative to its orbital plane, sunlight is incident at different angles at different times of the year. On Earth's surface, temperatures usually range ±40 °C (–40 °F to 100 °F) annually. Over thousands of years, changes in Earth's orbit can affect the amount and distribution of solar energy received by the Earth, thus influencing long-term climate and global climate change.

Surface temperature differences in turn cause pressure differences. Higher altitudes are cooler than lower altitudes due to differences in compressional heating. Weather forecasting is the application of science and technology to predict the state of the atmosphere for a future time and a given location.

The system is a chaotic system; so small changes to one part of the system can grow to have large effects on the system as a whole. Human attempts to control the weather have occurred throughout human history, and there is evidence that human activities such as agriculture and industry have modified weather patterns.

Studying how the weather works on other planets has been helpful in understanding how weather works on Earth. A famous landmark in the Solar System, Jupiter's *Great Red Spot*, is an anticyclonic storm known to have existed for at least 300 years. However, weather is not limited to planetary bodies. A star's corona is constantly being lost to space, creating what is essentially a very thin atmosphere throughout the Solar System. The movement of mass ejected from the Sun is known as the solar wind.

Causes

On Earth, the common weather phenomena include wind, cloud, rain, snow, fog and dust storms. Less common events include natural disasters such as tornadoes, hurricanes, typhoons and ice storms. Almost all familiar weather phenomena occur in the troposphere (the lower part of the atmosphere). Weather does occur in the stratosphere and can affect weather lower down in the troposphere, but the exact mechanisms are poorly understood.

Weather occurs primarily due to air pressure (temperature and moisture) differences between one place to another. These differences can occur due to the sun angle at any particular spot, which varies by latitude from the tropics.

In other words, the farther from the tropics one lies, the lower the sun angle is, which causes those locations to be cooler due to the indirect sunlight. The strong temperature contrast between polar and tropical air gives rise to the jet stream. Weather systems in the mid-latitudes, such as extratropical cyclones, are caused by instabilities of the jet stream flow. Weather systems in the tropics, such as monsoons or organized thunderstorm systems, are caused by different processes.

Because the Earth's axis is tilted relative to its orbital plane, sunlight is incident at different angles at different times of the year. In June the Northern Hemisphere is tilted towards the sun, so at any given Northern Hemisphere latitude sunlight falls more directly on that spot than in December. This effect causes seasons. Over thousands to hundreds of thousands of years, changes in Earth's orbital parametres affect the amount and distribution of solar energy received by the Earth and influence long-term climate.

The uneven solar heating (the formation of zones of temperature and moisture gradients, or frontogenesis) can also be due to the weather itself in the form of cloudiness and precipitation. Higher altitudes are cooler than lower altitudes, which is explained by the lapse rate. On local scales, temperature differences can occur because different surfaces (such as oceans, forests, ice sheets, or man-made objects) have differing physical characteristics such as reflectivity, roughness, or moisture content.

Surface temperature differences in turn cause pressure differences. A hot surface heats the air above it raising the pressure of the air (Combined gas law) and causing it to expand which then lowers the air pressure and its density. The resulting horizontal pressure gradient accelerates the air from high to low pressure, creating wind, and Earth's rotation then causes curvature of the flow via the Coriolis effect. The simple systems thus formed can then display emergent behaviour to produce more complex systems and thus other weather phenomena. Large scale examples include the Hadley cell while a smaller scale example would be coastal breezes.

The atmosphere is a chaotic system, so small changes to one part of the system can grow to have large effects on the system as a whole. This makes it difficult to accurately predict weather more than a few days in advance, though weather forecasters are continually working to extend this limit through the scientific study of weather, meteorology. It is theoretically impossible to make useful day-to-day predictions more than about two weeks ahead, imposing an upper limit to potential for improved prediction skill.

Shaping the Planet Earth

Weather is one of the fundamental processes that shape the Earth. The process of weathering breaks down the rocks and soils into smaller fragments and then into their constituent substances. During rains precipitation, the water droplets absorb and dissolve carbon dioxide from the surrounding air. This causes the rainwater to be slightly acidic, which aids the erosive properties of water.

The released sediment and chemicals are then free to take part in chemical reactions that can affect the surface further (such as acid rain), and sodium and chloride ions (salt) deposited in the seas/oceans. The sediment may reform in time and by geological forces into other rocks and soils. In this way, weather plays a major role in erosion of the surface.

Major Wind and Pressure Systems and Related Weather

Region	*Name*	*Pressure*	*Surface Winds*	*Weather*
Equator (0°)	Doldrums (ITCZ) (equatorial low)	Low	Light, variable winds	Cloudiness, abundant precipitation in all seasons; breeding ground for hurricanes. Relatively low sea-surface salinity because of high rainfall relative to evaporation
0°-30°N and S	Trade winds (easterlies)	-	Northeast in Northern Hemisphere; Southeast in Southern Hemisphere	Summer wet, winter dry; pathway for tropical disturbances
30°N and S	Horse latitudes	High	Light, variable winds	Little cloudiness; dry in all seasons. Relatively high sea-surface salinity because of high evaporation relative to precipitation
30°-60°N and S	Prevailing Westerlies	-	Southwest in Northern Hemisphere; Northwest in Southern Hemisphere	Winter wet, summer dry; pathway for subtropical high and low pressure
60°N and S	Polar front	Low	Variable	Stormy, cloudy weather zone; ample precipitation in all seasons
60°-90°N and S	Polar easterlies	-	Northeast in Northern Hemisphere; Southeast in Southern Hemisphere	Cold polar air with very low temperatures
90°N and S	Poles	High	Southerly in Northern Hemisphere; Northerly in Southern Hemisphere	Cold, dry air; sparse precipitation in all seasons

Effect on Humans

Effects on Populations: Weather has played a large and sometimes direct part in human history. Aside from climatic changes that have caused the gradual drift of populations (for example the desertification of the Middle East, and the formation of land bridges during glacial periods), extreme weather events have caused smaller scale population movements and intruded directly in historical events.

One such event is the saving of Japan from invasion by the Mongol fleet of Kublai Khan by the Kamikaze winds in 1281. French claims to Florida came to an end in 1565 when a hurricane destroyed the French fleet, allowing Spain to conquer Fort Caroline. More recently, Hurricane Katrina redistributed over one million people from the central Gulf coast elsewhere across the United States, becoming the largest diaspora in the history of the United States.

The Little Ice Age caused crop failures and famines in Europe. The 1690s saw the worst famine in France since the Middle Ages. Finland suffered a severe famine in 1696–1697, during which about one-third of the Finnish population died.

Forecasting

Weather forecasting is the application of science and technology to predict the state of the atmosphere for a future time and a given location. Human beings have attempted to predict the weather informally for millennia, and formally since at least the nineteenth century. Weather forecasts are made by collecting quantitative data about the current state of the atmosphere and using scientific understanding of atmospheric processes to project how the atmosphere will evolve.

Once an all-human endeavour based mainly upon changes in barometric pressure, current weather conditions, and sky condition, forecast models are now used to determine future conditions. Human input is still required to pick the best possible forecast model to base the forecast upon, which involves pattern recognition skills, teleconnections, knowledge of model performance, and knowledge of model biases. The chaotic nature of the atmosphere, the massive computational power required to solve the equations that describe the atmosphere, error involved in measuring the initial conditions, and an incomplete understanding of atmospheric processes mean that forecasts become less accurate as the difference in current time and the time for which the forecast is being made (the *range* of the forecast) increases. The use of ensembles and model consensus helps to narrow the error and pick the most likely outcome.

There are a variety of end users to weather forecasts. Weather warnings are important forecasts because they are used to protect life and property. Forecasts based on temperature and precipitation are important to agriculture, and therefore to commodity traders within stock markets. Temperature forecasts are used by utility companies to estimate demand over coming days. On an everyday basis, people use weather forecasts to determine what to wear on a given day. Since outdoor activities are severely curtailed by heavy rain, snow and the wind chill, forecasts can be used to plan activities around these events, and to plan ahead and survive them.

Modification

The aspiration to control the weather is evident throughout human history: from ancient rituals intended to bring rain for crops to the

U.S. Military Operation Popeye, an attempt to disrupt supply lines by lengthening the North Vietnamese monsoon. The most successful attempts at influencing weather involve cloud seeding; they include the fog- and low stratus dispersion techniques employed by major airports, techniques used to increase winter precipitation over mountains, and techniques to suppress hail.

A recent example of weather control was China's preparation for the 2008 Summer Olympic Games. China shot 1,104 rain dispersal rockets from 21 sites in the city of Beijing in an effort to keep rain away from the opening ceremony of the games on 8 August 2008. Guo Hu, head of the Beijing Municipal Meteorological Bureau (BMB), confirmed the success of the operation with 100 millimetres falling in Baoding City of Hebei Province, to the southwest and Beijing's Fangshan District recording a rainfall of 25 millimetres.

Whereas there is inconclusive evidence for these techniques' efficacy, there is extensive evidence that human activity such as agriculture and industry results in inadvertent weather modification:

- Acid rain, caused by industrial emission of sulphur dioxide and nitrogen oxides into the atmosphere, adversely affects freshwater lakes, vegetation, and structures.
- Anthropogenic pollutants reduce air quality and visibility.
- Climate change caused by human activities that emit greenhouse gases into the air is expected to affect the frequency of extreme weather events such as drought, extreme temperatures, flooding, high winds, and severe storms. However, some experts argue these claims are unfounded and take issue with these conclusions.
- Heat, generated by large metropolitan areas have been shown to minutely affect nearby weather, even at distances as far as 1,600 kilometres (990 mi).

The effects of inadvertent weather modification may pose serious threats to many aspects of civilization, including ecosystems, natural resources, food and fibre production, economic development, and human health.

Microscale Meteorology

Microscale meteorology is the study of short-lived atmospheric phenomena smaller than mesoscale, about 1 km or less. These two branches of meteorology are sometimes grouped together as "mesoscale and microscale meteorology" (MMM) and together study all phenomena smaller than synoptic scale; that is they study features generally too small to be depicted on a weather map. These include small and generally fleeting cloud "puffs" and other small cloud features.

Extremes on Earth

On Earth, temperatures usually range ±40 °C (100 °F to –40 °F) annually. The range of climates and latitudes across the planet can offer extremes of temperature outside this range. The coldest air temperature ever recorded on Earth is –89.2 °C (–128.6 °F), at Vostok Station, Antarctica on 21 July 1983.

The hottest air temperature ever recorded was 57.7 °C (135.9 °F) at 'Aziziya, Libya, on 13 September 1922, but that reading is queried. The highest recorded average annual temperature was 34.4 °C (93.9 °F) at Dallol, Ethiopia. The coldest recorded average annual temperature was –55.1 °C (–67.2 °F) at Vostok Station, Antarctica.

The coldest average annual temperature in a permanently inhabited location is at Eureka, Nunavut, in Canada, where the annual average temperature is –19.7 °C (–3.5 °F).

Extraterrestrial Within the Solar System

Studying how the weather works on other planets has been seen as helpful in understanding how it works on Earth. Weather on other planets follows many of the same physical principles as weather on Earth, but occurs on different scales and in atmospheres having different chemical composition. The Cassini–Huygens mission to Titan discovered clouds formed from methane or ethane which deposit rain composed of liquid methane and other organic compounds. Earth's atmosphere includes six latitudinal circulation zones, three in each hemisphere. In contrast, Jupiter's banded appearance shows many such zones, Titan has a single jet stream near the 50th parallel north latitude, and Venus has a single jet near the equator.

One of the most famous landmarks in the Solar System, Jupiter's *Great Red Spot*, is an anticyclonic storm known to have existed for at least 300 years. On other gas giants, the lack of a surface allows the wind to reach enormous speeds: gusts of up to 600 metres per second (about 2,100 km/h or 1,300 mph) have been measured on the planet Neptune. This has created a puzzle for planetary scientists.

The weather is ultimately created by solar energy and the amount of energy received by Neptune is only about $^{1}D_{900}$ of that received by Earth, yet the intensity of weather phenomena on Neptune is far greater than on Earth. The strongest planetary winds discovered so far are on the extrasolar planet HD 189733 b, which is thought to have easterly winds moving at more than 9,600 kilometres per hour (6,000 mph).

Space Weather

Weather is not limited to planetary bodies. Like all stars, the sun's corona is constantly being lost to space, creating what is essentially a very thin atmosphere throughout the Solar System. The movement of mass ejected from the Sun is known as the solar wind. Inconsistencies in this wind and larger events on the surface of the star, such as coronal mass ejections, form a system that has features analogous to conventional weather systems (such as pressure and wind) and is generally known as space weather. Coronal mass ejections have been tracked as far out in the solar system as Saturn.

The activity of this system can affect planetary atmospheres and occasionally surfaces. The interaction of the solar wind with the terrestrial atmosphere can produce spectacular aurorae, and can play havoc with electrically sensitive systems such as electricity grids and radio signals.

Climate

Climate is a measure of the average pattern of variation in temperature, humidity, atmospheric pressure, wind, precipitation, atmospheric particle count and other meteorological variables in a given region over long periods of time. Climate is different from weather, in that weather only describes the short-term conditions of these variables in a given region.

A region's climate is generated by the climate system, which has five components: atmosphere, hydrosphere, cryosphere, lithosphere, and biosphere.

The climate of a location is affected by its latitude, terrain, and altitude, as well as nearby water bodies and their currents. Climates can be classified according to the average and the typical ranges of different variables, most commonly temperature and precipitation. The most commonly used classification scheme was originally developed by Wladimir Köppen. The Thornthwaite system, in use since 1948, incorporates evapotranspiration along with temperature and precipitation information and is used in studying animal species diversity and potential effects of climate changes. The Bergeron and Spatial Synoptic Classification systems focus on the origin of air masses that define the climate of a region.

Paleoclimatology is the study of ancient climates. Since direct observations of climate are not available before the 19th century, paleoclimates are inferred from *proxy variables* that include non-biotic evidence such as sediments found in lake beds and ice cores, and biotic evidence such as tree rings and coral. Climate models are mathematical models of past, present and future climates. Climate change may occur over long and short timescales from a variety of factors; recent warming is discussed in global warming.

Climate Classification

There are several ways to classify climates into similar regimes. Originally, climes were defined in Ancient Greece to describe the weather depending upon a location's latitude. Modern climate classification methods can be broadly divided into *genetic* methods, which focus on the causes of climate, and *empiric* methods, which focus on the effects of climate. Examples of genetic classification include methods based on the relative frequency of different air mass types or locations within synoptic weather disturbances.

Examples of empiric classifications include climate zones defined by plant hardiness, evapotranspiration, or more generally the Köppen climate classification which was originally designed to identify the climates associated with certain biomes. A common shortcoming of these classification schemes is that they produce distinct boundaries between the zones they define, rather than the gradual transition of climate properties more common in nature.

Bergeron and Spatial Synoptic

The simplest classification is that involving air masses. The Bergeron classification is the most widely accepted form of air mass classification. Air mass classification involves three letters. The first letter describes its moisture properties, with c used for continental air masses (dry) and m for maritime air masses (moist). The second letter describes the thermal characteristic of its source region: T for tropical, P for polar, A for Arctic or Antarctic, M for monsoon, E for equatorial, and S for superior air (dry air formed by significant downward motion in the atmosphere).

The third letter is used to designate the stability of the atmosphere. If the air mass is colder than the ground below it, it is labelled k. If the air mass is warmer than the ground below it, it is labelled w. While air mass identification was originally used in weather forecasting during the 1950s, climatologists began to establish synoptic climatologies based on this idea in 1973.

Based upon the Bergeron classification scheme is the Spatial Synoptic Classification system (SSC). There are six categories within the SSC scheme: Dry Polar (similar to continental polar), Dry Moderate (similar to maritime superior), Dry Tropical (similar to continental tropical), Moist Polar (similar to maritime polar), Moist Moderate (a hybrid between maritime polar and maritime tropical), and Moist Tropical (similar to maritime tropical, maritime monsoon, or maritime equatorial).

Köppen

The Köppen classification depends on average monthly values of temperature and precipitation. The most commonly used form of the Köppen classification has five primary types labelled A through E. These primary types are A, tropical; B, dry; C, mild mid-latitude; D, cold mid-latitude; and E, polar. The five primary classifications can be further divided into secondary classifications such as rain forest, monsoon, tropical savanna, humid subtropical, humid continental, oceanic climate, Mediterranean climate, steppe, subarctic climate, tundra, polar ice cap, and desert.

Rain forests are characterized by high rainfall, with definitions setting minimum normal annual rainfall between 1,750 millimetres (69 in) and 2,000 millimetres (79 in). Mean monthly temperatures exceed 18 °C (64 °F) during all months of the year.

A monsoon is a seasonal prevailing wind which lasts for several months, ushering in a region's rainy season. Regions within North America, South America, Sub-Saharan Africa, Australia and East Asia are monsoon regimes.

A tropical savanna is a grassland biome located in semiarid to semi-humid climate regions of subtropical and tropical latitudes, with average temperatures remain at or above 18 °C (64 °F) year round and rainfall between 750 millimetres (30 in) and 1,270 millimetres (50 in) a year. They are widespread on Africa, and are found in India, the northern parts of South America, Malaysia, and Australia.

The humid subtropical climate zone where winter rainfall (and sometimes snowfall) is associated with large storms that the westerlies steer from west to east. Most summer rainfall occurs during thunderstorms and from occasional tropical cyclones. Humid subtropical climates lie on the east side continents, roughly between latitudes 20° and 40° degrees away from the equator.

A humid continental climate is marked by variable weather patterns and a large seasonal temperature variance. Places with more than three months of average daily temperatures above 10 °C (50 °F) and a coldest month temperature below –3 °C (27 °F) and which do not meet the criteria for an arid or semiarid climate, are classified as continental.

An oceanic climate is typically found along the west coasts at the middle latitudes of all the world's continents, and in southeastern Australia, and is accompanied by plentiful precipitation year round.

The Mediterranean climate regime resembles the climate of the lands in the Mediterranean Basin, parts of western North America, parts of Western and South Australia, in southwestern South Africa and in parts of central Chile. The climate is characterized by hot, dry summers and cool, wet winters.

A steppe is a dry grassland with an annual temperature range in the summer of up to 40 °C (104 °F) and during the winter down to –40 °C (–40 °F).

A subarctic climate has little precipitation, and monthly temperatures which are above 10 °C (50 °F) for one to three months of the year, with permafrost in large parts of the area due to the cold winters. Winters within subarctic climates usually include up to six months of temperatures averaging below 0 °C (32 °F).

Tundra occurs in the far Northern Hemisphere, north of the taiga belt, including vast areas of northern Russia and Canada.

A polar ice cap, or polar ice sheet, is a high-latitude region of a planet or moon that is covered in ice. Ice caps form because high-latitude regions receive less energy as solar radiation from the sun than equatorial regions, resulting in lower surface temperatures.

A desert is a landscape form or region that receives very little precipitation. Deserts usually have a large diurnal and seasonal temperature range, with high or low, depending on location daytime temperatures (in summer up to 45 °C or 113 °F), and low nighttime temperatures (in winter down to 0 °C or 32 °F) due to extremely low humidity. Many deserts are formed by rain shadows, as mountains block the path of moisture and precipitation to the desert.

Thornthwaite

Devised by the American climatologist and geographer C. W. Thornthwaite, this climate classification method monitors the soil water budget using evapotranspiration. It monitors the portion of total precipitation used to nourish vegetation over a certain area. It uses indices such as a humidity index and an aridity index to determine an area's moisture regime based upon its average temperature, average rainfall, and average vegetation type. The lower the value of the index in any given area, the drier the area is.

The moisture classification includes climatic classes with descriptors such as hyperhumid, humid, subhumid, subarid, semi-arid (values of –20 to –40), and arid (values below –40). Humid regions experience more precipitation than evaporation each year, while arid regions experience greater evaporation than precipitation on an annual basis.

A total of 33 percent of the Earth's landmass is considered either arid of semi-arid, including southwest North America, southwest South America, most of northern and a small part of southern Africa, southwest and portions of eastern Asia, as well as much of Australia. Studies suggest that precipitation effectiveness (PE) within the Thornthwaite moisture index is overestimated in the summer and underestimated in the winter.

This index can be effectively used to determine the number of herbivore and mammal species numbers within a given area. The index is also used in studies of climate change.

Thermal classifications within the Thornthwaite scheme include microthermal, mesothermal, and megathermal regimes. A microthermal climate is one of low annual mean temperatures, generally between 0 °C (32 °F) and 14 °C (57 °F) which experiences short summers and has a potential evaporation between 14 centimetres (5.5 in) and 43 centimetres (17 in).

A mesothermal climate lacks persistent heat or persistent cold, with potential evaporation between 57 centimetres (22 in) and 114 centimetres (45 in). A megathermal climate is one with persistent high temperatures and abundant rainfall, with potential annual evaporation in excess of 114 centimetres (45 in). For example climate in Nepal was very hot while the climate in United States was very cold.

Record

Modern: Details of the modern climate record are known through the taking of measurements from such weather instruments as thermometres, barometres, and anemometres during the past few centuries. The instruments used to study weather over the modern time scale, their known error, their immediate environment, and their exposure have changed over the years, which must be considered when studying the climate of centuries past.

Paleoclimatology

Paleoclimatology is the study of past climate over a great period of the Earth's history. It uses evidence from ice sheets, tree rings, sediments, coral, and rocks to determine the past state of the climate. It demonstrates periods of stability and periods of change and can indicate whether changes follow patterns such as regular cycles.

***Table:** Difference between weather and climate*

Sr No	*Weather*	*Climate*
1	Instantaneous physical state of atmosphere at particular place.	Normal physical state or generated condition of atmosphere or long term average condition of a place.
2	Weather changes refer to specific instant of time(day or week)	It is generalized over a longer span of time and for a longer area.
3	It is expressed in terms of numerical values of meteorological elements.	It is expressed in terms of time averages and area averages of meteorological elements.
4	Weather is measured in observatory. So the observatory must at a place for which weather is to be described.	This is derived information on regional basis. So scripts of observatories extending over a region are necessary.
5	No statistical treatment is applied to the meteorological elements. They are used as observed and hence always changing.	Application of statistical method over a longer period I done. It is more or less stable with few random changes.
6	it provides meteorological information.	It constitutes geographical information in respect of weather.
7	Weather of two places having same numerical value must be same.	Climate of the two places having the same averages of weather can not be same, because their distribution over the years may be different.
8	Weather can be categorized as fair, unfair, excellent etc.	Climate is classified as desert climate, marine climate, tropical climate etc.
9	Weather decides the success or failure of a crop in a particular season.	Climate decides the type of crop suitable for a region, while introducing new crops climate is considered.
10	Adverse weather results into crop failure or loss and warrants short term contingent planning.	Climate is considered in long terms agricultural planning.

Chapter 2

Climatic Controls

The value of weather elements are modified by the interference of the factors of determining causes like latitude, altitude, etc. Such factors are called as climatic factors of climatic controls.

Latitude

In geography, latitude (φ) is a geographic coordinate that specifies the north-south position of a point on the Earth's surface. Latitude is an angle (defined below) which ranges from 0° at the Equator to 90° (North or South) at the poles. Lines of constant latitude, or parallels, run east–west as circles parallel to the equator. Latitude is used together with longitude to specify the precise location of features on the surface of the Earth. Two levels of abstraction are employed in the definition of these coordinates. In the first step the physical surface is modelled by the geoid, a surface which approximates the mean sea level over the oceans and its continuation under the land masses. The second step is to approximate the geoid by a mathematically simpler reference surface. The simplest choice for the reference surface is a sphere, but the geoid is more accurately modelled by an ellipsoid. The definitions of latitude and longitude on such reference surfaces. Lines of constant latitude and longitude together constitute a graticule on the reference surface. The latitude of a point on the *actual* surface is that of the corresponding point on the reference surface, the correspondence being along the normal to the reference surface which passes through the point on the physical surface. Latitude and longitude together with some specification of height constitute a geographic coordinate system as defined in the specification of the ISO 19111 standard.

Since there are many different reference ellipsoids the latitude of a feature on the surface is not unique: this is stressed in the ISO standard which states that "without the full specification of the coordinate reference system, coordinates (that is latitude and longitude) are ambiguous at best and meaningless at worst". This is of great importance in accurate applications, such as GPS, but in common usage, where high accuracy is not required, the reference ellipsoid is not usually stated.

In English texts the latitude angle, defined below, is usually denoted by the Greek lower-case letter phi (φ or x). It is measured in degrees, minutes and seconds or decimal degrees, north or south of the equator. Measurement of latitude requires an understanding of the gravitational field of the Earth, either for setting up theodolites or for determination of GPS satellite orbits. The study of the figure of the Earth together with its gravitational field is the science of geodesy. These topics are not discussed in this article.

This article relates to coordinate systems for the Earth: it may be extended to cover the Moon, planets and other celestial objects by a simple change of nomenclature.

Latitude effects climate depending on how close you are to the equator. If you are close to the equator then the climate is warmer, if your farther away then it is colder, in general. The reason for this is that the closer you are to the equator, the more directly the sun's rays go through the Earth's atmosphere. This rule is complicated by many things, including ocean currents and weather patterns, but in general: the closer you are to the equator, the warmer it will be.

Altitude

Altitude or height is defined based on the context in which it is used (aviation, geometry, geographical survey, sport, and more). As a general definition, altitude is a distance measurement, usually in the vertical or "up" direction, between a reference datum and a point or object. The reference datum also often varies according to the context. Although the term altitude is commonly used to mean the height above sea level of a location, in geography the term elevation is often preferred for this usage.

Vertical distance measurements in the "down" direction are commonly referred to as depth.

Lapse Rate

The lapse rate is defined as the rate at which atmospheric temperature decreases with increase in altitude. The terminology arises from the word *lapse* in the sense of a decrease or decline. While most often applied to Earth's troposphere, the concept can be extended to any gravitationally supported ball of gas.

Definition

A formal definition from the *Glossary of Meteorology* is: The decrease of an atmospheric variable with height, the variable being temperature unless otherwise specified.

In the lower regions of the atmosphere (up to altitudes of approximately 12,000 metres (39,000 ft), temperature decreases with altitude at a fairly uniform rate. Because the atmosphere is warmed by convection from Earth's surface, this lapse or reduction in temperature is normal with increasing distance from the conductive source. Although the actual atmospheric lapse rate varies, under normal atmospheric conditions the average atmospheric lapse rate results in a temperature decrease of 6.4C°/km (3.5F°/1,000 ft) of altitude above ground level.

The measurable lapse rate is affected by the moisture content of the air (humidity). A dry lapse rate of 10C°/km (5.5F°/1,000 ft) is often used to calculate temperature changes in air not at 100% relative humidity. A wet lapse rate of 5.5C°/km (3F°/1,000 ft) is used to calculate the temperature changes in air that is saturated (i.e., air at 100% relative humidity). Although actual lapse rates do not strictly follow these guidelines, they present a model sufficiently accurate to predict temperate changes associated with updrafts and downdrafts. This differential lapse rate (dependent upon both difference in conductive heating and adiabatic expansion or compression) results in the formation of warm downslope winds (e.g., Chinook winds, Santa Ana winds, etc.). The atmospheric lapse rate, combined with adiabatic cooling and heating of air related to the expansion and compression of atmospheric gases, present a unified model explaining the cooling of air as it moves aloft and the heating of air as it descends downslope.

Atmospheric stability can be measured in terms of lapse rates (i.e., the temperature differences associated with vertical movement of air). The atmosphere is considered conditionally unstable where the environmental lapse rate causes a slower decrease in temperature with altitude than the dry adiabatic lapse rate, as long as no latent heat is released (i.e. the saturated adiabatic lapse rate applies). Unconditional instability results when the dry adiabatic lapse rate causes air to cool slower than the environmental lapse rate, so air will continue to rise until it reaches the same temperature as its surroundings. Where the saturated adiabatic lapse rate is greater than the environmental lapse rate, the air cools faster than its environment and thus returns to its original position, irrespective of its moisture content. Although the atmospheric lapse rate (also known as the environmental lapse rate) is most often used to characterize temperature changes, many properties (e.g. atmospheric pressure) can also be profiled by lapse rates...

Mathematical Definition

In general, a lapse rate is the negative of the rate of temperature change with altitude change, thus:

$$\gamma = -\frac{dT}{dz}$$

where γ is the lapse rate given in units of temperature divided by units of altitude, T = temperature, and z = altitude.

Note: In some cases, Γ or α can be used to represent the adiabatic lapse rate in order to avoid confusion with other terms symbolized by, such as the specific heat ratio or the psychrometric constant.

Types of Lapse Rates

There are two types of lapse rate:

- Environmental lapse rate – which refers to the actual change of temperature with altitude for the stationary atmosphere (i.e. the temperature gradient)
- The adiabatic lapse rates – which refer to the change in temperature of a parcel of air as it moves upwards (or downwards) without exchanging heat with its surroundings. The temperature change that occurs within the air parcel reflects the adjusting balance between potential energy and kinetic energy of the molecules of gas that comprise the moving air mass. There are two adiabatic rates:
    - o Dry adiabatic lapse rate
    - o Moist (or saturated) adiabatic lapse rate

Environmental Lapse Rate

The environmental lapse rate (ELR), is the rate of decrease of temperature with altitude in the stationary atmosphere at a given time and location. As an average, the International Civil Aviation Organization (ICAO) defines an international standard atmosphere (ISA) with a temperature lapse rate of 6.49 K(C°)/1,000 m (3.56 F° or 1.98 K(C°)/1,000 Ft) from sea level to 11 km (36,090 ft). From 11 km (36,090 ft or 6.8 mi) up to 20 km (65,620 ft or 12.4 mi), the constant temperature is –56.5 C° (–69.7 F°), which is the lowest assumed temperature in the ISA. The standard atmosphere contains no moisture. Unlike the idealized ISA, the temperature of the actual atmosphere does not always fall at a uniform rate with height. For example, there can be an inversion layer in which the temperature increases with altitude.

Dry Adiabatic Lapse Rate

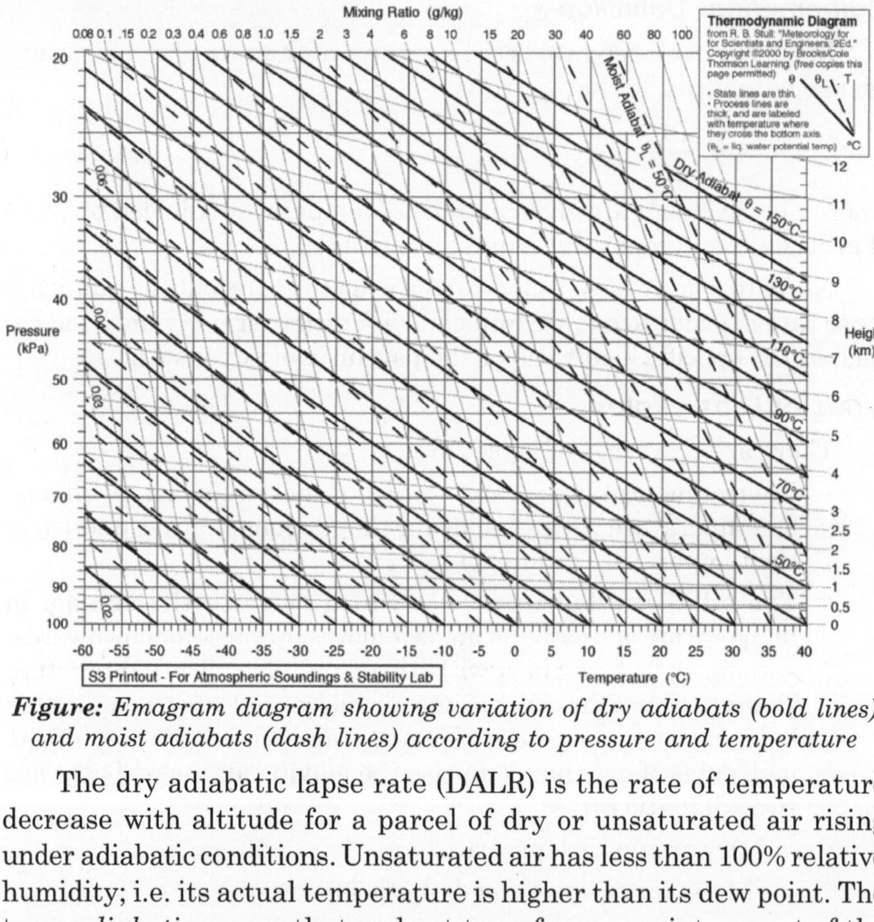

Figure: *Emagram diagram showing variation of dry adiabats (bold lines) and moist adiabats (dash lines) according to pressure and temperature*

The dry adiabatic lapse rate (DALR) is the rate of temperature decrease with altitude for a parcel of dry or unsaturated air rising under adiabatic conditions. Unsaturated air has less than 100% relative humidity; i.e. its actual temperature is higher than its dew point. The term *adiabatic* means that no heat transfer occurs into or out of the parcel. Air has low thermal conductivity, and the bodies of air involved are very large, so transfer of heat by conduction is negligibly small.

Under these conditions when the air rises (for instance, by convection) it expands, because the pressure is lower at higher altitudes. As the air parcel expands, it pushes on the air around it, doing work (thermodynamics). Since the parcel does work but gains no heat, it loses internal energy so that its temperature decreases. The rate of temperature decrease is 9.8 C°/km (5.38 F° per 1,000 ft) (3.0C°/1,000 ft). The reverse occurs for a sinking parcel of air.

Since for adiabatic process:

$$PdV = -VdP / \gamma$$

the first law of thermodynamics can be written as

$$mc_v dT - Vdp / \gamma = 0$$

Also since : $\alpha = V / m$ and : $\gamma = c_p / c_v$ we can show that:

$$c_p dT - \alpha dP = 0$$

where c_p is the specific heat at constant pressure and α is the specific volume.

Assuming an atmosphere in hydrostatic equilibrium:

$$dP = -\rho g dz$$

where g is the standard gravity and ρ is the density. Combining these two equations to eliminate the pressure, one arrives at the result for the DALR,

$$\Gamma_d = -\frac{dT}{dz} = \frac{g}{c_p} = 9.8\ ^{\circ}\mathrm{C/km}$$

Saturated Adiabatic Lapse Rate

When the air is saturated with water vapour (at its dew point), the moist adiabatic lapse rate (MALR) or saturated adiabatic lapse rate (SALR) applies. This lapse rate varies strongly with temperature. A typical value is around 5 C°/km (2.7 F°/1,000 ft) (1.5C°/1,000 ft).

The reason for the difference between the dry and moist adiabatic lapse rate values is that latent heat is released when water condenses, thus decreasing the rate of temperature drop as altitude increases. This heat release process is an important source of energy in the development of thunderstorms.

An unsaturated parcel of air of given temperature, altitude and moisture content below that of the corresponding dewpoint cools at the *dry adiabatic lapse rate* as altitude increases until the dewpoint line for the given moisture content is intersected. As the water vapour then starts condensing the air parcel subsequently cools at the slower *moist adiabatic lapse rate* if the altitude increases further.

The saturated adiabatic lapse rate is given approximately by this equation from the glossary of the American Meteorology Society:

$$\Gamma_w = g\frac{1+\dfrac{H_v r}{R_{sd}T}}{c_{pd}+\dfrac{H_v^2 r}{R_{sw}T^2}} = g\frac{1+\dfrac{H_v r}{R_{sd}T}}{c_{pd}+\dfrac{H_v^2 r\epsilon}{R_{sd}T^2}}$$

	where:
Γ_w	= Wet adiabatic lapse rate, K/m
g	= Earth's gravitational acceleration = 9.8076 m/s^2
H_v	= Heat of vaporization of water, = 2501000 J/kg
R_{sd}	= Specific gas constant of dry air = 287 J kg^{-1} K^{-1}
R_{sw}	= Specific gas constant of water vapour = 461.5 J kg^{-1} K^{-1}
$\epsilon = \dfrac{R_{sd}}{R_{sw}}$	=The dimensionless ratio of the specific gas constant of dry air to the specific gas constant for water vapour = 0.622
e	= The water vapour pressure of the saturated air
p	= The pressure of the saturated air
$r = \epsilon e/(p-e)$	= The mixing ratio of the mass of water vapour to the mass of dry air
T	= Temperature of the saturated air, K
c_{pd}	= The specific heat of dry air at constant pressure, = 1003.5 J kg^{-1} K^{-1}

Thermodynamic-based Lapse Rate

Robert Essenhigh developed a comprehensive thermodynamic model of the lapse rate based on the Schuster–Schwarzschild (S–S) integral equations of transfer that govern radiation through the atmosphere including absorption and radiation by greenhouse gases. His solution "predicts, in agreement with the Standard Atmosphere experimental data, a linear decline of the fourth power of the temperature, T^4, with pressure, P, and, as a first approximation, a linear decline of T with altitude, h, up to the tropopause at about 10 km (the lower atmosphere)." The predicted normalized density ratio and pressure ratio differ and fit the experimental data well. Sreekanth Kolan extended Essenhigh's model to include the energy balance for the lower and upper atmospheres.

Significance in Meteorology

The varying environmental lapse rates throughout the Earth's atmosphere are of critical importance in meteorology, particularly within the troposphere. They are used to determine if the parcel of rising air will rise high enough for its water to condense to form clouds, and, having formed clouds, whether the air will continue to rise and form bigger shower clouds, and whether these clouds will get even bigger and form cumulonimbus clouds (thunder clouds).

As unsaturated air rises, its temperature drops at the dry adiabatic rate. The dew point also drops (as a result of decreasing air pressure) but much more slowly, typically about –2 C° per 1,000 m. If unsaturated air rises far enough, eventually its temperature will reach its dew point, and condensation will begin to form. This altitude is known as the lifting condensation level (LCL) when mechanical lift is present and the convective condensation level (CCL) when mechanical lift is absent, in which case, the parcel must be heated from below to its convective temperature. The cloud base will be somewhere within the layer bounded by these parametres.

The difference between the dry adiabatic lapse rate and the rate at which the dew point drops is around 8C° per 1,000 m. Given a difference in temperature and dew point readings on the ground, one can easily find the LCL by multiplying the difference by 125 m/C°.

If the environmental lapse rate is less than the moist adiabatic lapse rate, the air is absolutely stable — rising air will cool faster than the surrounding air and lose buoyancy. This often happens in the early morning, when the air near the ground has cooled overnight. Cloud formation in stable air is unlikely.

If the environmental lapse rate is between the moist and dry adiabatic lapse rates, the air is conditionally unstable — an unsaturated parcel of air does not have sufficient buoyancy to rise to the LCL or CCL, and it is stable to weak vertical displacements in either direction. If the parcel is saturated it is unstable and will rise to the LCL or CCL, and either be halted due to an inversion layer of convective inhibition, or if lifting continues, deep, moist convection (DMC) may ensue, as a parcel rises to the level of free convection (LFC), after which it enters the free convective layer (FCL) and usually rises to the equilibrium level (EL).

If the environmental lapse rate is larger than the dry adiabatic lapse rate, it has a superadiabatic lapse rate, the air is absolutely unstable — a parcel of air will gain buoyancy as it rises both below and above the lifting condensation level or convective condensation level. This often happens in the afternoon over many land masses. In these conditions, the likelihood of cumulus clouds, showers or even thunderstorms is increased.

Meteorologists use radiosondes to measure the environmental lapse rate and compare it to the predicted adiabatic lapse rate to forecast the likelihood that air will rise. Charts of the environmental lapse rate are known as thermodynamic diagrams, examples of which include Skew-T log-P diagrams and tephigrams.

The difference in moist adiabatic lapse rate and the dry rate is the cause of foehn wind phenomenon (also known as "Chinook winds" in parts of North America).

Topography

Topography is a field of geoscience and planetary science comprising the study of surface shape and features of the Earth and other observable astronomical objects including planets, moons, and asteroids. It is also the description of such surface shapes and features (especially their depiction in maps). The topography of an area could also mean the surface shape and features themselves.

In general, topography is concerned with local detail in general, including not only relief but also natural and artificial features, and even local history and culture. This meaning is less common in America, where topographic maps with elevation contours have made "topography" synonymous with relief. The older sense of topography as the study of place still has currency in Europe.

Topography in a narrow sense involves the recording of relief or terrain, the three-dimensional quality of the surface, and the identification of specific landforms. This is also known as geomorphometry. In modern usage, this involves generation of elevation data in digital form (DEM). It is often considered to include the graphic representation of the landform on a map by a variety of techniques, including contour lines, hypsometric tints, and relief shading.

Wind velocity primarily changes with change in topography which may result in Change in temperature.

Mountain

A mountain is a large landform that stretches above the surrounding land in a limited area, usually in the form of a peak. A mountain is generally steeper than a hill. Mountains are formed through tectonic forces or volcanism. These forces can locally raise the surface of the earth. Mountains erode slowly through the action of rivers, weather conditions, and glaciers. A few mountains are isolated summits, but most occur in huge mountain ranges.

High elevations on mountains produce colder climates than at sea level. These colder climates strongly affect the ecosystem of mountains: different elevations have different plants and animals. Because of the less hospitable terrain and climate, mountains tend to be used less for agriculture and more for resource extraction and recreation, such as mountain climbing. The highest mountain on Earth is Mount Everest in the Himalayas of Asia, whose summit is 8,850 m (29,035 ft) above mean sea level. The highest known mountain on any planet in the Solar System is Olympus Mons on Mars at 21,171 m (69,459 ft).

Alpine Climate

Alpine climate is the average weather (climate) for a region above the tree line. This climate is also referred to as a mountain climate or highland climate.

In the Köppen climate classification, the alpine climate is part of "Group E", along with the polar climate, where no month has a mean temperature higher than 10 °C (50 °F). Certain highland climates can also fit under the hemiboreal climate or semi-arid climate groups of climate classification.

Description

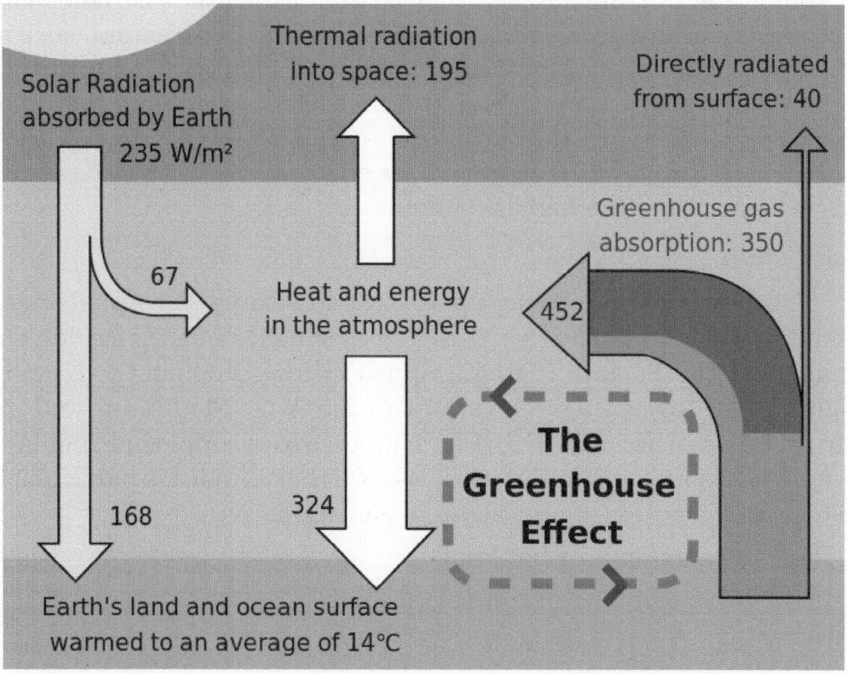

Figure: *Diagram showing heat transfer in the Earth's atmosphere*

The climate becomes colder at high elevations, due to the way that the sun heats the surface of the Earth. Practically all the heat at the surface of the Earth comes from the sun, in the form of solar energy. The sun's radiation is absorbed by land and sea, which is warmed. The warm land loses heat by convection within the atmosphere, and long-wave radiation back to space.

This radiation can move freely through gases composed of diatomic molecules, such as the atmosphere's oxygen and nitrogen, but is readily absorbed and re-radiated by triatomic molecules, such as carbon dioxide and water vapour. When the heat is re-radiated, some of the heat that would be lost to space is instead reflected back towards the Earth. Thus, the troposphere, as a whole, acts as a blanket for the Earth. This blanket effect is known as the "greenhouse effect". The higher the altitude, the less of this blanket there is to keep in the heat. Thus, higher elevations, such as mountains, are colder than surrounding lowlands.

The rate at which the temperature drops with elevation, called the environmental lapse rate, is not constant (it can fluctuate throughout the day or seasonally and also regionally), but a normal lapse rate is 5.5°C per 1,000 m (3.57°F per 1,000 ft). Therefore, moving up 100 metres (330 ft) on a mountain is roughly equivalent to moving 80 kilometres (45 miles or 0.75° of latitude) towards the pole. This relationship is only approximate, however, since local factors, such as proximity to oceans, can drastically modify the climate. As the altitude increases, the main form of precipitation becomes snow and the winds increase. The temperature continues to drop until the tropopause, at 11,000 metres (36,000 ft), where it does not decrease further. However, this is higher than the highest summit.

Distribution

The Cascade Mountains, the Rocky Mountains, the Appalachian Mountains, the Alps, the Pyrenees and Sierra Nevada, the Andes, the Himalayas, the Tibetan Plateau, Gansu China, Qinghai, the Eastern Highlands of Africa, high elevations in the Atlas Mountains and the central parts of Borneo and New Guinea are examples of highland climates. The mountain climate in the Northern Andes is particularly known for the notion of four zones of elevation:

- *Tierra caliente* or hot land
- *Tierra templada* or temperate land
- *Tierra fría* or cold land
- *Tierra helada* or frozen land

In mountainous areas with an alpine climate, the dominant biome is alpine tundra.

Land and Sea Distribution

Distribution of land and sea has a profound effect on climate. Places near the sea have moderate climate. On the other hand places for away from the sea are very hot in summer and very cold in winter. So they are said to have an extreme climate.

Climates around the world are shaped by the transfer of heat energy via the general circulation of the atmosphere. The idealised model of climate zones, pressure patterns and global wind belts that can be sketched however, is complicated by the position of continents and oceans. Land surfaces react quickly to heat gain and loss, becoming warm in summer, cold in winter. The oceans react far more slowly and during the summer they are cooler than the adjoining land, whilst in winter they are warmer.

High atmospheric pressure usually develops over cold areas were air is descending, whilst low pressure forms over warmer regions where air is rising. During the Northern Hemisphere winter a large high-pressure anticyclone develops over Asia, centred on Siberia where temperatures can fall to -40°C. A weaker winter anticyclone develops over North America. In contrast, low-pressure centres known as depressions continue to develop in the warmer North Atlantic and North Pacific Oceans, centred respectively on Iceland and the Aleutian Islands. These are the Icelandic and Aleutian Lows.

In summer, the landmasses warm up, and the winter high pressure over Asia is replaced by a large low-pressure system, centred over northern India. The Icelandic and Aleutian Lows weaken and the subtropical highs become stronger. The build up of low pressure over Asia assists the development of the monsoon over the Indian Ocean, where the more usual northeast trade winds are replaced by southwesterlies which bring moist air and a prolonged wet season to India and Southeast Asia.

Similar modifications to pressure and wind patterns occur in the Southern Hemisphere, although these are less pronounced than in the Northern Hemisphere because of the absence of really large landmasses.

The moderating influence of the oceans helps to restrict extremes in climate in coastal areas of the world. The typical annual average temperature range in the UK is only about 10 to 15°C, whereas in central Siberia or central Canada it can be over 40°C. For this reason maritime climates influenced by airflow from the oceans are usually more pleasant than the continental climates of landmass interiors, although they are frequently wetter.

Oceans

The significance of the oceans is that they store a much greater quantity of energy than the atmosphere. Heat transfers between the oceans and the atmosphere have a major influence on climates. In addition, the oceans are major reservoirs of moisture, and coastal zones in particular frequently experience humid and wet conditions that are common to maritime climates.

As in the atmosphere, energy in the oceans is circulated around the world, transferring heat from low latitudes to high latitudes. This energy is redistributed by a global system of ocean currents interconnected to form the global ocean circulation. Surface ocean currents are driven by the atmospheric winds. Deep ocean currents are generated by the sinking and upwelling of water, due to density differences driven by temperature and salinity (saltiness) contrasts.

A well known and important surface ocean current is the Gulf Stream, which originates in the Gulf of Mexico and flows northeast across the Atlantic, driven by the prevailing southwest winds. The Gulf Stream is a warm ocean current that keeps the climates of UK and Northwest Europe mild in winter.

Unlike the Gulf Stream, which is a permanent feature of the global ocean circulation, the El Niño phenomenon develops roughly every 2 to 7 years, and usually lasts for several months, although on occasion has prevailed for several years. El Niño is the name given to this sporadic development of warm surface waters in equatorial Pacific Ocean. El Niño has a huge influence on regional climate, bringing with in heavy rainfall and flooding to the west coast of South America, and drought and bush fires to parts of Australasia.

During strong El Niños, its influence may even extend globally. 1998, for example, became the warmest year on record, in large part due to the presence of a strong El Niño.

How the Oceans Influence Climate

The oceans influence climate over long and short time-scales. On the longest time-scale of geologic time, the shape and location of the continents helps to determine the oceans' circulation patterns. Since continental plates drift at about 5 cm per year and mountain ranges rise by about 1 mm, it usually takes millions of years for new land formations to change the oceans. Patterns of ocean circulation and upwelling can also change much more rapidly, resulting in climate variations and fluctuations on a human time-scale.

Records of global and, in particular, regional climate show periods lasting from years to centuries during which the climate was systematically different from earlier and later periods. Scientists believe that this behaviour is related to changes in the way the oceans store and transport heat, although the precise causes of these changes are not always clear.

The oceans and the atmosphere are tightly linked and together form the most dynamic component of the climate system. Changes in external factors such the sun's energy, the distribution of various plant species, or the emission of greenhouse gases into the atmosphere can alter the temperature and circulation patterns of the atmosphere-ocean system. Because the atmosphere and oceans are turbulent, they can also generate their own internal fluctuations.

Short-term fluctuations in wind or temperature (that is, weather) can directly influence the currents and temperature of the underlying ocean, while oceanic fluctuations can magnify, diminish, or modify atmospheric fluctuations.

The oceans play a critical role in storing heat and carbon). When the earth's surface cools or is heated by the sun, the temperature change is greater - and faster - over the land than over the oceans. Because it is a fluid, the ocean diffuses the effects of a temperature change for great distances via vertical mixing and convective movements.

The solid land cannot, so the sun's heat penetrates only the thin, upper crust. One consequence of the ocean's ability to absorb more heat is that when an area of ocean becomes warmer or cooler than usual, it takes much longer for that area to revert to "normal" than it would for a land area. This also explains why "maritime" climates tend to be less extreme than "continental" ones, with smaller day-night and winter-summer differences.

The ocean's waters are constantly being moved about by powerful currents. Surface currents are largely wind-driven, although the rotation of the earth, the presence of continents, and the oceans' internal dynamics also have a strong influence. Deep-ocean flow (and, to a lesser extent, surface flow) is driven by density differences produced by heating and cooling and by precipitation and evaporation (cool saline water is denser than warm fresh-water). The behaviour of the atmosphere strongly affects these density differences. For example, clouds can cool the sea by blocking the warming rays of the sun or reduce surface salinity by bringing rain. The wind can influence evaporation rates by blowing more strongly or more weakly.

These currents influence the climate by transporting heat. Horizontal currents, particularly those moving north or south, can carry warmed or cooled water as far as several thousand kilometres. The displaced water can then warm or cool the air and, indirectly, the land over which this air blows.

For example, water from the tropical and subtropical Atlantic (including some from the Gulf of Mexico) moves north through the Atlantic in a current popularly (if misleadingly) called the "Gulf Stream". There it bathes the shores of Western Europe, producing a climate that is surprisingly mild for that latitude. In addition to currents, up-wellings of cold water in places where the wind blows surface water away can also affect climate.

Thus San Francisco, influenced by coastal up-welling, is hardly warmer than Dublin, which is influenced by the Gulf Stream, despite being over 1,600 km further south.

Currents involved in "deep-water formation" are particularly important for climate. In winter, surface cooling causes water to become more dense. (While fresh-water that is cooled starts to expand at temperatures below 4 C, salt-water continues to compress all the way down to its freezing point of -2 C.) In areas where evaporation exceeds precipitation, the resulting rise in salinity also increases density. When the surface water becomes denser than the underlying water, "convective overturning" occurs and the dense surface water mixes downwards. In certain places this downward mixing can occasionally extend all the way to the bottom, even in deep oceans.

The dense, deep water thus formed spreads out over the whole ocean. As a result, when downward mixing takes place at high latitudes it creates a circulation pattern in which warm water from tropical and subtropical regions moves poleward, surrenders heat to the atmosphere, cools and sinks, and flows back towards the equator. The net result is a transport of heat poleward.

An apparently small change in just one aspect of the ocean's behaviour can produce major climate variations over large areas of the earth. The areas of cold-water formation are one known example of this possibly wide-spread phenomenon. Although more research is needed, there is some agreement among oceanographers that, for the entire area north of 30 N latitude, the ocean's poleward transport of heat is the equivalent of about 15 watts per square metre of the earth's surface (W/m2). This can be compared with some 200 W/m2 from direct sunshine, and about 6 W/m2 for what climate change models predict will happen if the atmospheric concentration of carbon dioxide doubles.

Recent observations, ocean core records, and some modelling results indicate that North Atlantic deep-water formation and its associated ocean heat flow fluctuate substantially over time-scales ranging from years to millennia. The system is vulnerable because even a relatively small decrease in surface salinity prevents water - no matter how cold it is - from sinking. This could occur if there is a flood of fresh-water run-off from the Arctic due to global warming.

Chapter 3

Atmosphere of Earth

The atmosphere of Earth is a layer of gases surrounding the planet Earth that is retained by Earth's gravity. The atmosphere protects life on Earth by absorbing ultraviolet solar radiation, warming the surface through heat retention (greenhouse effect), and reducing temperature extremes between day and night (the diurnal temperature variation).

The common name given to the atmospheric gases used in breathing and photosynthesis is air. By volume, dry air contains 78.09% nitrogen, 20.95% oxygen, 0.93% argon, 0.039% carbon dioxide, and small amounts of other gases. Air also contains a variable amount of water vapour, on average around 1%. Although air content and atmospheric pressure vary at different layers, air suitable for the survival of terrestrial plants and terrestrial animals currently is only known to be found in Earth's troposphere and artificial atmospheres.

The atmosphere has a mass of about 5.15×10^{18} kg, three quarters of which is within about 11 km (6.8 mi; 36,000 ft) of the surface. The atmosphere becomes thinner and thinner with increasing altitude, with no definite boundary between the atmosphere and outer space. The Kármán line, at 100 km (62 mi), or 1.57% of Earth's radius, is often used as the border between the atmosphere and outer space. Atmospheric effects become noticeable during atmospheric reentry of spacecraft at an altitude of around 120 km (75 mi). Several layers can be distinguished in the atmosphere, based on characteristics such as temperature and composition.

The study of Earth's atmosphere and its processes is called atmospheric science or aerology. Early pioneers in the field include Léon Teisserenc de Bort and Richard Assmann.

Composition

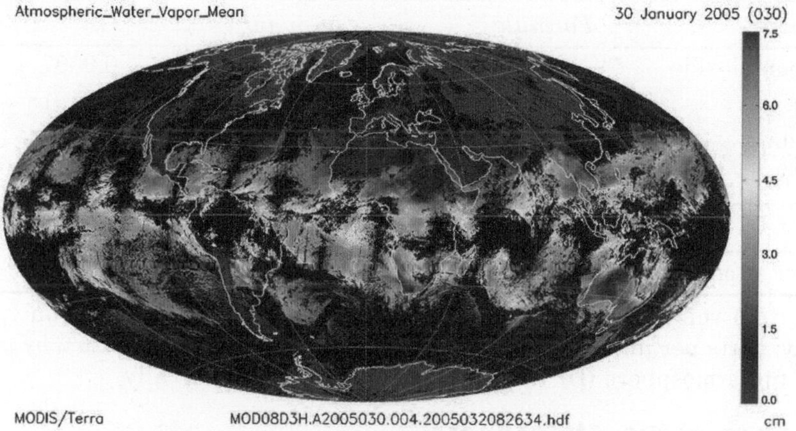

Fgure: *Mean atmospheric water vapour*

Air is mainly composed of nitrogen, oxygen, and argon, which together constitute the major gases of the atmosphere. Water vapour accounts for roughly 0.25% of the atmosphere by mass. The concentration of water vapour (a greenhouse gas) varies significantly from around 10 ppmv in the coldest portions of the atmosphere to as much as 5% by volume in hot, humid air masses, and concentrations of other atmospheric gases are typically provided for dry air without any water vapour. The remaining gases are often referred to as trace gases, among which are the greenhouse gases such as carbon dioxide, methane, nitrous oxide, and ozone. Filtered air includes trace amounts of many other chemical compounds. Many substances of natural origin may be present in locally and seasonally variable small amounts as aerosols in an unfiltered air sample, including dust of mineral and organic composition, pollen and spores, sea spray, and volcanic ash. Various industrial pollutants also may be present as gases or aerosols, such as chlorine (elemental or in compounds), fluorine compounds and elemental mercury vapour. Sulphur compounds such as hydrogen sulfide and sulphur dioxide (SO_2) may be derived from natural sources or from industrial air pollution.

Major constituents of dry air, by volume

Gas		*Volume*[(A)]	
Name	*Formula*	*in ppmv*[(B)]	*in %*
Nitrogen	N_2	780,840	78.084
Oxygen	O_2	209,460	20.946
Argon	Ar	9,340	0.9340

Contd.

Name	*Formula*	*in ppmv*[(B)]	*in %*
Carbon dioxide	CO2	397	0.0397
Neon	Ne	18.18	0.001818
Helium	He	5.24	0.000524
Methane	CH_4	1.79	0.000179
Not included in above dry atmosphere:			
Water vapor[(C)]	H_2O	10–50,000[(D)]	0.001%–5%[(D)]

Notes: (A) volume fraction is equal to mole fraction for ideal gas only, (B) ppmv: parts per million by volume (C) Water vapour is about 0.25% *by mass* over full atmosphere (D) Water vapour strongly varies locally

Structure of the Atmosphere

Principal Layers

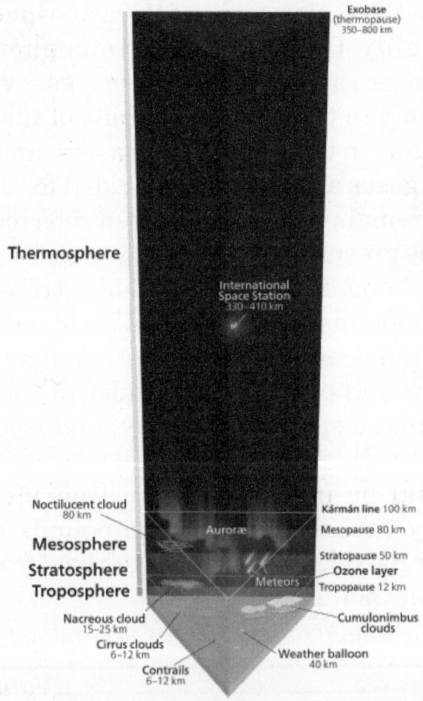

Figure: *Earth's atmosphere Lower 4 layers of the atmosphere in 3 dimensions as seen diagonally from above the exobase. Layers drawn to scale, objects within the layers are not to scale. Aurorae shown here at the bottom of the thermosphere can actually form at any altitude in this atmospheric layer*

In general, air pressure and density decrease with altitude in the atmosphere. However, temperature has a more complicated profile with altitude, and may remain relatively constant or even increase with altitude in some regions. Because the general pattern of the temperature/altitude profile is constant and recognizable through means such as balloon soundings, the temperature behaviour provides a useful metric to distinguish between atmospheric layers. In this way, Earth's atmosphere can be divided (called atmospheric stratification) into five main layers. From highest to lowest, these layers are:

- Exosphere: >700 km (>440 miles)
- Thermosphere: 80 to 700 km (50 to 440 miles)
- Mesosphere: 50 to 80 km (31 to 50 miles)
- Stratosphere: 12 to 50 km (7 to 31 miles)
- Troposphere: 0 to 12 km (0 to 7 miles)

Exosphere

The exosphere is the outermost layer of Earth's atmosphere (i.e. the upper limit of the atmosphere). It extends from the exobase, which is located at the top of the thermosphere at an altitude of about 700 km above sea level, to about 10,000 km (6,200 mi; 33,000,000 ft). The exosphere merges with the emptiness of outer space, where there is no atmosphere.

This layer is mainly composed of extremely low densities of hydrogen, helium and several heavier molecules including nitrogen, oxygen and carbon dioxide closer to the exobase. The atoms and molecules are so far apart that they can travel hundreds of kilometres without colliding with one another. Thus, the exosphere no longer behaves like a gas, and the particles constantly escape into space. These free-moving particles follow ballistic trajectories and may migrate in and out of the magnetosphere or the solar wind.

The exosphere is located too far above Earth for any meteorological phenomena to be possible. However, the aurora borealis and aurora australis sometimes occur in the lower part of the exosphere, where they overlap into the thermosphere. The exosphere contains most of the satellites orbiting Earth.

Thermosphere

The thermosphere is the second-highest layer of Earth's atmosphere. It extends from the mesopause (which separates it from the mesosphere) at an altitude of about 80 km (50 mi; 260,000 ft) up to the thermopause at an altitude range of 500–1000 km (310–620 mi;

1,600,000–3,300,000 ft). The height of the thermopause varies considerably due to changes in solar activity. Since the thermopause lies at the lower boundary of the exosphere, it is also referred to as the exobase. The lower part of the thermosphere, from 80 to 550 kilometres (50 to 342 mi) above Earth's surface, contains the ionosphere.

This atmospheric layer undergoes a gradual increase in temperature with height. Unlike the stratosphere, wherein a temperature inversion is due to the absorption of radiation by ozone, the inversion in the thermosphere occurs due to the extremely low density of its molecules.

The temperature of this layer can rise as high as 1500 °C (2700 °F), though the gas molecules are so far apart that its temperature in the usual sense is not very meaningful.

The air is so rarefied that an individual molecule (of oxygen, for example) travels an average of 1 kilometre (0.62 mi; 3300 ft) between collisions with other molecules. Even though the thermosphere has a very high proportion of molecules with immense amounts of energy, the thermosphere would still feel extremely cold to a human in direct contact because the total energy of its relatively few number of molecules is incapable of transferring an adequate amount of energy to the skin of a human. In other words, a person would not feel warm because of the thermosphere's extremely low pressure.

This layer is completely cloudless and free of water vapour. However non-hydrometeorological phenomena such as the aurora borealis and aurora australis are occasionally seen in the thermosphere. The International Space Station orbits in this layer, between 320 and 380 km (200 and 240 mi).

Mesosphere

The mesosphere is the third highest layer of Earth's atmosphere, occupying the region above the stratosphere and below the thermosphere. It extends from the stratopause at an altitude of about 50 km (31 mi; 160,000 ft) to the mesopause at 80–85 km (50–53 mi; 260,000–280,000 ft) above sea level.

Temperatures drop with increasing altitude to the mesopause that marks the top of this middle layer of the atmosphere. It is the coldest place on Earth and has an average temperature around –85 °C (–120 °F; 190 K).

Just below the mesopause, the air is so cold that even the very scarce water vapour at this altitude can be sublimated into polarmesospheric noctilucent clouds. These are highest clouds in the atmosphere and may be visible to the naked eye if sunlight reflects off them about an hour or two after sunset or a similar length of time before sunrise. They are most readily visible when the Sun is around 4 to 16 degrees below the horizon.

A type of lightning referred to as either sprites or ELVES, occasionally form far above tropospheric thunderclouds. The mesosphere is also the layer where most meteors burn up upon atmospheric entrance. It is too high above Earth to be accessible to jet-powered aircraft, and too low to support satellites and orbital or sub-orbital spacecraft. The mesosphere is mainly accessed by rocket-powered aircraft and unmanned sounding rockets.

Stratosphere

The stratosphere is the second-lowest layer of Earth's atmosphere. It lies above the troposphere and is separated from it by the tropopause. This layer extends from the top of the troposphere at roughly 12 km (7.5 mi; 39,000 ft) above Earth's surface to the stratopause at an altitude of about 50 to 55 km (31 to 34 mi; 164,000 to 180,000 ft).

The atmospheric pressure at the top of the stratosphere is roughly 1/1000 the pressure at sea level. It contains the ozone layer, which is the part of Earth's atmosphere that contains relatively high concentrations of that gas. The stratosphere defines a layer in which temperatures rise with increasing altitude. This rise in temperature is caused by the absorption of ultraviolet radiation (UV) radiation from the Sun by the ozone layer, which restricts turbulence and mixing. Although the temperature may be –60 °C (–76 °F; 210 K) at the tropopause, the top of the stratosphere is much warmer, and may be near 0 °C.

The stratospheric temperature profile creates very stable atmospheric conditions, so the stratosphere lacks the weather-producing air turbulence that is so prevalent in the troposphere. Consequently, the stratosphere is almost completely free of clouds and other forms of weather. However, polar stratospheric or nacreous clouds are occasionally seen in the lower part of this layer of the atmosphere where the air is coldest. This is the highest layer that can be accessed by jet-powered aircraft.

Troposphere

The troposphere is the lowest layer of Earth's atmosphere. It extends from Earth's surface to an average height of about 12 km, although this altitude actually varies from about 9 km (30,000 ft) at the poles to 17 km (56,000 ft) at the equator, with some variation due to weather. The troposphere is bounded above by the tropopause, a boundary marked by stable temperatures.

Although variations do occur, the temperature usually declines with increasing altitude in the troposphere because the troposphere is mostly heated through energy transfer from the surface. Thus, the lowest part of the troposphere (i.e. Earth's surface) is typically the warmest section of the troposphere.

The troposphere contains roughly 80% of the mass of Earth's atmosphere. The troposphere is denser than all its overlying atmospheric layers because a larger atmospheric weight sits on top of the troposphere and causes it to be most severely compressed. Fifty percent of the total mass of the atmosphere is located in the lower 5.6 km (18,000 ft) of the troposphere. It is primarily composed of nitrogen (78%) and oxygen (21%) with only small concentrations of other trace gases.

Nearly all atmospheric water vapour or moisture is found in the troposphere, so it is the layer where most of Earth's weather takes place. It has basically all the weather-associated cloud genus types generated by active wind circulation, although very tall cumulonimbus thunder clouds can penetrate the tropopause from below and rise into the lower part of the stratosphere.

Most conventional aviation activity takes place in the troposphere, and it is the only layer that can be accessed by propeller-driven aircraft.

Other Layers

Within the five principal layers which are largely determined by temperature, several secondary layers may be distinguished by other properties:

- The ozone layer is contained within the stratosphere. In this layer ozone concentrations are about 2 to 8 parts per million, which is much higher than in the lower atmosphere but still very small compared to the main components of the atmosphere. It is mainly located in the lower portion of the stratosphere from about 15–35 km (9.3–21.7 mi; 49,000–115,000 ft), though the thickness varies seasonally and geographically. About 90% of the ozone in our atmosphere is contained in the stratosphere.
- The ionosphere is a region of the atmosphere that is ionized by solar radiation. It is responsible for auroras. During daytime

hours, it stretches from 50 to 1,000 km (31 to 621 mi; 160,000 to 3,280,000 ft) and includes the mesosphere, thermosphere, and parts of the exosphere.

However, ionization in the mesosphere largely ceases during the night, so auroras are normally seen only in the thermosphere and lower exosphere. The ionosphere forms the inner edge of the magnetosphere. It has practical importance because it influences, for example, radio propagation on Earth.

- The homosphere and heterosphere are defined by whether the atmospheric gases are well mixed.

  The surfaced-based homosphere includes the troposphere, stratosphere, mesosphere, and the lowest part of the thermosphere, where the chemical composition of the atmosphere does not depend on molecular weight because the gases are mixed by turbulence.

  This relatively homogeneous layer ends at the *turbopause* which is found at about 100 km (62 mi; 330,000 ft), which places it about 20 km (12 mi; 66,000 ft) above the mesopause.

Above this altitude lies the heterosphere which includes the exosphere and most of the thermosphere. Here the chemical composition varies with altitude.

This is because the distance that particles can move without colliding with one another is large compared with the size of motions that cause mixing. This allows the gases to stratify by molecular weight, with the heavier ones such as oxygen and nitrogen present only near the bottom of the heterosphere.

The upper part of the heterosphere is composed almost completely of hydrogen, the lightest element.

- The planetary boundary layer is the part of the troposphere that is closest to Earth's surface and is directly affected by it, mainly through turbulent diffusion. During the day the planetary boundary layer usually is well-mixed, whereas at night it becomes stably stratified with weak or intermittent mixing. The depth of the planetary boundary layer ranges from as little as about 100 metres on clear, calm nights to 3000 m or more during the afternoon in dry regions.

The average temperature of the atmosphere at the surface of Earth is 14 °C (57 °F; 287 K) or 15 °C (59 °F; 288 K), depending on the reference.

Physical Properties

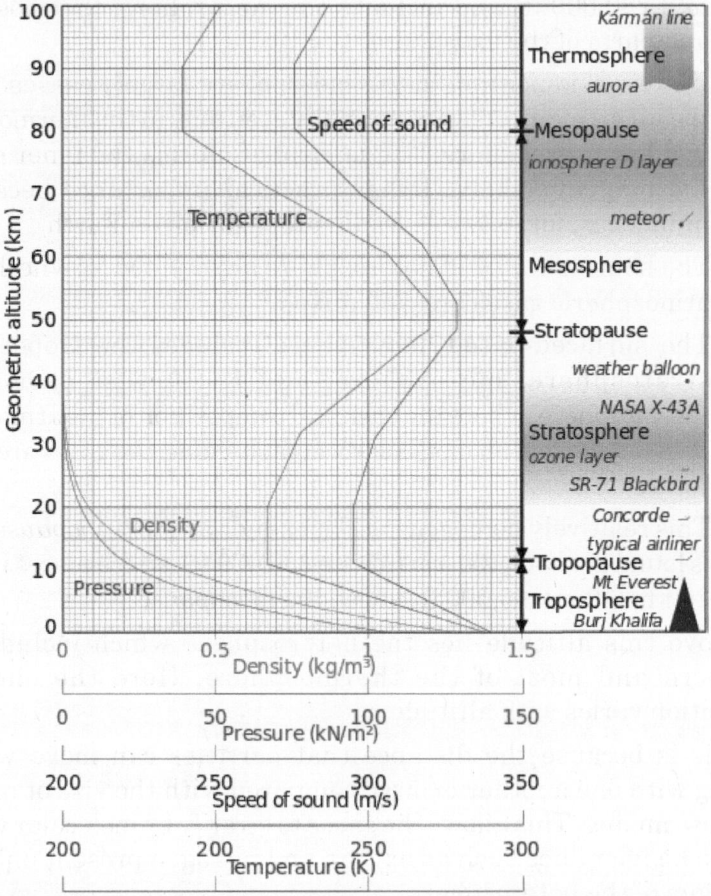

***Figure:** Comparison of the 1962 US Standard Atmosphere graph of geometric altitude against air density, pressure, the speed of sound and temperature with approximate altitudes of various objects.*

Pressure and Thickness

The average atmospheric pressure at sea level is defined by the International Standard Atmosphere as 101325 pascals (760.00 Torr; 14.6959 psi; 760.00 mmHg). This is sometimes referred to as a unit of standard atmospheres (atm). Total atmospheric mass is 5.1480×10^{18} kg (1.135×10^{19} lb), about 2.5% less than would be inferred from the average sea level pressure and Earth's area of 51007.2 megahectares, this portion being displaced by Earth's mountainous terrain. Atmospheric pressure is the total weight of the air above unit area at the point where the pressure is measured. Thus air pressure varies with location and weather.

If the entire mass of the atmosphere had a uniform density from sea level, it would terminate abruptly at an altitude of 8.50 km (27,900 ft). It actually decreases exponentially with altitude, dropping by half every 5.6 km (18,000 ft) or by a factor of 1/e every 7.64 km (25,100 ft), the average scale height of the atmosphere below 70 km (43 mi; 230,000 ft). However, the atmosphere is more accurately modelled with a customized equation for each layer that takes gradients of temperature, molecular composition, solar radiation and gravity into account.

In summary, the mass of Earth's atmosphere is distributed approximately as follows:

- 50% is below 5.6 km (18,000 ft).
- 90% is below 16 km (52,000 ft).
- 99.99997% is below 100 km (62 mi; 330,000 ft), the Kármán line. By international convention, this marks the beginning of space where human travellers are considered astronauts.

By comparison, the summit of Mt. Everest is at 8,848 m (29,029 ft); commercial airliners typically cruise between 10 km (33,000 ft) and 13 km (43,000 ft) where the thinner air improves fuel economy; weather balloons reach 30.4 km (100,000 ft) and above; and the highest X-15 flight in 1963 reached 108.0 km (354,300 ft).

Even above the Kármán line, significant atmospheric effects such as auroras still occur. Meteors begin to glow in this region though the larger ones may not burn up until they penetrate more deeply. The various layers of Earth's ionosphere, important to HF radio propagation, begin below 100 km and extend beyond 500 km. By comparison, the International Space Station and Space Shuttle typically orbit at 350–400 km, within the F-layer of the ionosphere where they encounter enough atmospheric drag to require reboosts every few months. Depending on solar activity, satellites can experience noticeable atmospheric drag at altitudes as high as 700–800 km.

Temperature and Speed of Sound

The division of the atmosphere into layers mostly by reference to temperature is discussed above. Temperature decreases with altitude starting at sea level, but variations in this trend begin above 11 km, where the temperature stabilizes through a large vertical distance through the rest of the troposphere. In the stratosphere, starting above about 20 km, the temperature increases with height, due to heating within the ozone layer caused by capture of significant ultraviolet radiation from the Sun by the dioxygen and ozone gas in this region.

Still another region of increasing temperature with altitude occurs at very high altitudes, in the aptly-named thermosphere above 90 km.

Because in an ideal gas of constant composition the speed of sound depends only on temperature and not on the gas pressure or density, the speed of sound in the atmosphere with altitude takes on the form of the complicated temperature profile, and does not mirror altitudinal changes in density or pressure.

Density and Mass

The density of air at sea level is about 1.2 kg/m^3 (1.2 g/L). Density is not measured directly but is calculated from measurements of temperature, pressure and humidity using the equation of state for air (a form of the ideal gas law). Atmospheric density decreases as the altitude increases. This variation can be approximately modelled using the barometric formula. More sophisticated models are used to predict orbital decay of satellites.

The average mass of the atmosphere is about 5 quadrillion (5×10^{15}) tonnes or 1/1,200,000 the mass of Earth. According to the American National Centre for Atmospheric Research, "The total mean mass of the atmosphere is 5.1480×10^{18} kg with an annual range due to water vapour of 1.2 or 1.5×10^{15} kg depending on whether surface pressure or water vapour data are used; somewhat smaller than the previous estimate. The mean mass of water vapour is estimated as 1.27×10^{16} kg and the dry air mass as $5.1352 \pm 0.0003\times10^{18}$ kg."

Optical Properties

Solar radiation (or sunlight) is the energy Earth receives from the Sun. Earth also emits radiation back into space, but at longer wavelengths that we cannot see. Part of the incoming and emitted radiation is absorbed or reflected by the atmosphere.

Scattering

When light passes through our atmosphere, photons interact with it through *scattering*. If the light does not interact with the atmosphere, it is called *direct radiation* and is what you see if you were to look directly at the Sun. *Indirect radiation* is light that has been scattered in the atmosphere. For example, on an overcast day when you cannot see your shadow there is no direct radiation reaching you, it has all been scattered. As another example, due to a phenomenon called Rayleigh scattering, shorter (blue) wavelengths scatter more easily than longer (red) wavelengths. This is why the sky looks blue; you are seeing scattered blue light. This is also why sunsets are red. Because the Sun is close to the horizon, the Sun's rays pass through more atmosphere than normal to reach your eye. Much of the blue light has been scattered out, leaving the red light in a sunset.

Absorption

Different molecules absorb different wavelengths of radiation. For example, O_2 and O_3 absorb almost all wavelengths shorter than 300 manometres. Water (H_2O) absorbs many wavelengths above 700 nm. When a molecule absorbs a photon, it increases the energy of the molecule. This heats the atmosphere, but the atmosphere also cools by emitting radiation, as discussed below.

The combined absorption spectra of the gases in the atmosphere leave "windows" of low opacity, allowing the transmission of only certain bands of light.

The optical window runs from around 300 nm (ultraviolet-C) up into the range humans can see, the visible spectrum (commonly called light), at roughly 400–700 nm and continues to the infrared to around 1100 nm.

There are also infrared and radio windows that transmit some infrared and radio waves at longer wavelengths. For example, the radio window runs from about one centimetre to about eleven-metre waves.

Emission

Emission is the opposite of absorption, it is when an object emits radiation. Objects tend to emit amounts and wavelengths of radiation depending on their "black body" emission curves, therefore hotter objects tend to emit more radiation, with shorter wavelengths. Colder objects emit less radiation, with longer wavelengths. For example, the Sun is approximately 6,000 K (5,730 °C; 10,340 °F), its radiation peaks near 500 nm, and is visible to the human eye. Earth is approximately 290 K (17 °C; 62 °F), so its radiation peaks near 10,000 nm, and is much too long to be visible to humans.

Because of its temperature, the atmosphere emits infrared radiation. For example, on clear nights Earth's surface cools down faster than on cloudy nights. This is because clouds (H_2O) are strong absorbers and emitters of infrared radiation. This is also why it becomes colder at night at higher elevations.

The greenhouse effect is directly related to this absorption and emission effect. Some gases in the atmosphere absorb and emit infrared radiation, but do not interact with sunlight in the visible spectrum. Common examples of these are CO_2 and H_2O.

Refractive Index

The refractive index of air is close to, but just greater than 1. Systematic variations in refractive index can lead to the bending of light rays over long optical paths. One example is that, under some circumstances, observers onboard ships can see other vessels just over the horizon because light is refracted in the same direction as the curvature of Earth's surface.

The refractive index of air depends on temperature, giving rise to refraction effects when the temperature gradient is large. An example of such effects is the mirage.

Circulation

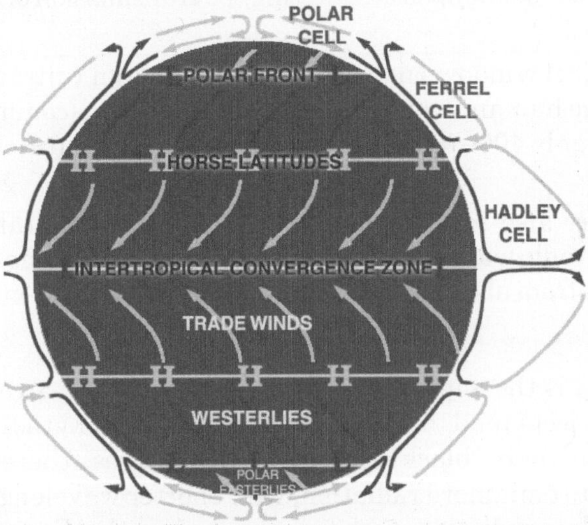

Figure: *An idealised view of three large circulation cells.*

Atmospheric circulation is the large-scale movement of air through the troposphere, and the means (with ocean circulation) by which heat is distributed around Earth. The large-scale structure of the atmospheric circulation varies from year to year, but the basic structure remains fairly constant because it is determined by Earth's rotation rate and the difference in solar radiation between the equator and poles.

Evolution of Earth's Atmosphere

Earliest Atmosphere

The first atmosphere would have consisted of gases in the solar nebula, primarily hydrogen. In addition, there would probably have been simple hydrides such as those now found in gas giants like Jupiter and Saturn, notably water vapour, methane and ammonia. As the solar nebula dissipated, these gases would have escaped, partly driven off by the solar wind.

Second Atmosphere

The next atmosphere, consisting largely of nitrogen plus carbon dioxide and inert gases, was produced by outgassing from volcanism, supplemented by gases produced during the late heavy bombardment of Earth by huge asteroids. A major part of carbon dioxide emissions were soon dissolved in water and built up carbonate sediments.

Water-related sediments have been found dating from as early as 3.8 billion years ago. About 3.4 billion years ago, nitrogen was the major part of the then stable "second atmosphere". An influence of life has to be taken into account rather soon in the history of the atmosphere, because hints of early life forms are to be found as early as 3.5 billion years ago. How Earth at that time managed a climate warm enough for liquid water and life, if the early Sun put out 30% lower solar radiance than today, is a puzzle known as the "faint young Sun paradox".

The geological record however shows a continually relatively warm surface during the complete early temperature record of Earth with the exception of one cold glacial phase about 2.4 billion years ago. In the late Archaean eon an oxygen-containing atmosphere began to develop, apparently produced by photosynthesizing cyanobacteria which have been found as stromatolite fossils from 2.7 billion years ago. The early basic carbon isotopy (isotope ratio proportions) is very much in line with what is found today, suggesting that the fundamental features of the carbon cycle were established as early as 4 billion years ago.

Earth's dynamic oxygenation evolution is recorded in ancient sediments from the Republic of Gabon from between about 2,150 and 2,080 million years ago. These fluctuations in oxygenation were likely driven by the Lomagundi carbon isotope excursion.

Third Atmosphere

The constant re-arrangement of continents by plate tectonics influences the long-term evolution of the atmosphere by transferring carbon dioxide to and from large continental carbonate stores. Free oxygen did not exist in the atmosphere until about 2.4 billion years ago during the Great Oxygenation Event and its appearance is indicated by the end of the banded iron formations. Before this time, any oxygen produced by photosynthesis was consumed by oxidation of reduced materials, notably iron. Molecules of free oxygen did not start to accumulate in the atmosphere until the rate of production of oxygen began to exceed the availability of reducing materials.

This point signifies a shift from a reducing atmosphere to an oxidizing atmosphere. O_2 showed major variations until reaching a steady state of more than 15% by the end of the Precambrian. The following time span from 541 million years ago to the present day is the Phanerozoic eon, during the earliest period of which, the Cambrian, oxygen-requiring metazoan life forms began to appear.

The amount of oxygen in the atmosphere has fluctuated over the last 600 million years, reaching a peak of about 30% around 280 million years ago, significantly higher than today's 21%. Two main processes govern changes in the atmosphere: Plants use carbon dioxide from the atmosphere, releasing oxygen. Breakdown of pyrite and volcanic eruptions release sulphur into the atmosphere, which oxidizes and hence reduces the amount of oxygen in the atmosphere.

However, volcanic eruptions also release carbon dioxide, which plants can convert to oxygen. The exact cause of the variation of the amount of oxygen in the atmosphere is not known. Periods with much oxygen in the atmosphere are associated with rapid development of animals. Today's atmosphere contains 21% oxygen, which is high enough for this rapid development of animals.

Currently, anthropogenic greenhouse gases are accumulating in the atmosphere, which is the main cause of global warming.

Air Pollution

Air pollution is the introduction into the atmosphere of chemicals, particulate matter, or biological materials that cause harm or discomfort to organisms. Stratospheric ozone depletion is believed to be caused by air pollution (chiefly from chlorofluorocarbons).

Aerosol

An aerosol is a colloid of fine solid particles or liquid droplets, in air or another gas. Examples of aerosols include haze, dust, particulate air pollutants and smoke. The liquid or solid particles have diametre mostly smaller than 1 μm or so; larger particles with a significant settling speed make the mixture a suspension, but the distinction is not clear-cut. In general conversation, *aerosol* usually refers to an aerosol spray that delivers a consumer product from a can or similar container. Other technological applications of aerosols include dispersal of pesticides, medical treatment of respiratory illnesses, and combustion technology. Diseases can also spread by means of small droplets in the breath, also called aerosols.

Aerosol science covers generation and removal of aerosols, technological application of aerosols, effects of aerosols on the environment and people, and a wide variety of other topics.

Definitions

An aerosol is defined as a colloidal system of solid or liquid particles in a gas. An aerosol includes both the particles and the suspending gas, which is usually air. Frederick G. Donnan presumably first used the term *aerosol* during World War I to describe an aero-solution, clouds of microscopic particles in air. This term developed analogously to the term hydrosol, a colloid system with water as the dispersing medium. *Primary aerosols* contain particles introduced directly into the gas; *secondary aerosols* form through gas-to-particle conversion. There are several measures of aerosol concentration. Environmental science and health often uses the *mass concentration* (*M*), defined as the mass of particulate matter per unit volume with units such as $\mu g/m^3$. Also commonly used is the *number concentration* (*N*), the number of particles per unit volume with units such as number/m^3 or number/cm^3.

The size of particles has a major influence on their properties, and the aerosol particle radius or diametre (d_p) is a key property used to characterise aerosols.

Aerosols vary in their dispersity. A *monodisperse* aerosol, producible in the laboratory, contains particles of uniform size. Most aerosols, however, as *polydisperse* colloidal systems, exhibit a range of particle sizes. Liquid droplets are almost always nearly spherical, but scientists use an *equivalent diametre* to characterize the properities of various shapes of solid particles, some very irregular. The equivalent diametre is the diametre of a spherical particle with the same value of some physical property as the irregular particle. The *equivalent volume diametre* (d_e) is defined as the diametre of a sphere of the same volume as that of the irregular particle. Also commonly used is the aerodynamic diametre.

Size Distribution

For a monodisperse aerosol, a single number—the particle diametre—suffices to describe the size of the particles. However, more complicated particle-size distributions describe the sizes of the particles in a polydisperse aerosol. This distribution defines the relative amounts of particles, sorted according to size.

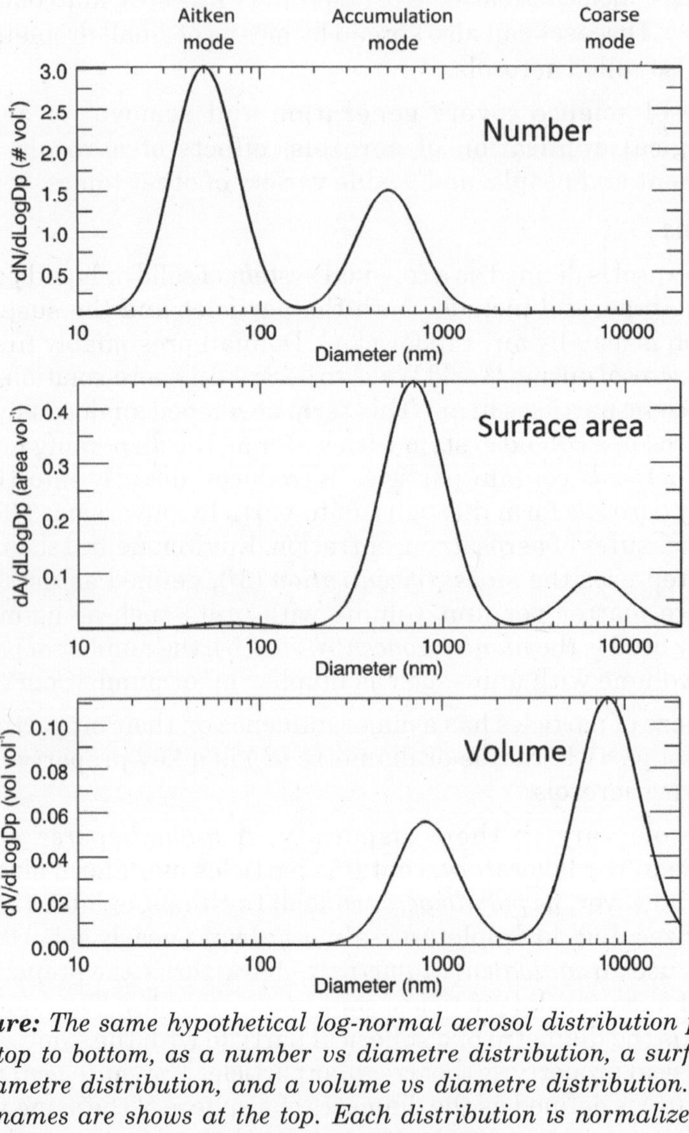

***Figure:** The same hypothetical log-normal aerosol distribution plotted, from top to bottom, as a number vs diametre distribution, a surface area vs diametre distribution, and a volume vs diametre distribution. Typical mode names are shows at the top. Each distribution is normalized so that the total area is 1000.*

One approach to defining the particle size distribution uses a list of the sizes of every particle in a sample. However, this approach proves tedious to ascertain in aerosols with millions of particles and awkward to use. Another approach splits the complete size range into intervals and finds the number (or proportion) of particles in each interval. One then can visualize these data in a histogram with the area of each bar representing the proportion of particles in that size bin, usually normalised by dividing the number of particles in a bin by the width of the interval so that the area of each bar is proportionate to the number of particles in the size range that it represents. If the width of the bins tends to zero, one gets the frequency function:

$$df = f(d_p)dd_p$$

where

d_p is the diametre of the particles

df is the fraction of particles having diametres between d_p and d_p + dd_p $f(d_p)$ is the frequency function

Therefore, the area under the frequency curve between two sizes a and *b* represents the total fraction of the particles in that size range:

$$f_{ab} = \int_a^b f(d_p)dd_p$$

It can also be formulated in terms of the total number density *N*:

$$dN = N(d_p)dd_p$$

Assuming spherical aerosol particles, the aerosol surface area per unit volume (*S*) is given by the second moment:

$$S = \pi / 2 \int_0^\infty N(d_p)d_p^2 dd_p$$

And the third moment gives the total volume concentration (*V*) of the particles:

$$V = \pi / 6 \int_0^\infty N(d_p)d_p^3 dd_p$$

One also usefully can approximate the particle size distribution using a mathematical function. The normal distribution usually does not suitably describe particle size distributions in aerosols because of the skewness associated a long tail of larger particles. Also for a quantity that varies over a large range, as many aerosol sizes do, the width of the distribution implies negative particles sizes, clearly not physically realistic. However, the normal distribution can be suitable for some aerosols, such as test aerosols, certain pollen grains and spores.

A more widely chosen log-normal distribution gives the number frequency as:

$$df = \frac{1}{\sigma\sqrt{2\pi}} e^{-\frac{(d_p - \overline{d_p})^2}{2\sigma^2}} dd_p$$

where:

σ is the standard deviation of the size distribution and

$\overline{d_p}$ is the arithmetic mean diametre.

The log-normal distribution has no negative values, can cover a wide range of values, and fits many observed size distributions reasonably well.

Other distributions sometimes used to characterise particle size include: the Rosin-Rammler distribution, applied to coarsely dispersed dusts and sprays; the Nukiyama-Tanasawa distribution, for sprays of extremely broad size ranges; the power function distribution, occasionally applied to atmospheric aerosols; the exponential distribution, applied to powdered materials; and for cloud droplets, the Khrgian-Mazin distribution.

Physics

Terminal Velocity of a Particle in a Fluid

For low values of the Reynolds number (<1), true for most aerosol motion, Stokes' law describes the force of resistance on a solid spherical particle in a fluid. However, Stokes' law is only valid when the velocity of the gas at the surface of the particle is zero. For small particles (< 1 μm) that characterize aerosols, however, this assumption fails. To account for this failure, one can introduce the Cunningham correction factor, always greater than 1. Including this factor, one finds the relation between the resisting force on a particle and its velocity:

$$F_D = \frac{3\pi\eta V d}{C_c}$$

where

F_D is the resisting force on a spherical particle

η is the viscosity of the gas

V is the particle velocity

C_c is the Cunningham correction factor.

This allows us to calculate the terminal velocity of a particle undergoing gravitational settling in still air. Neglecting buoyancy effects, we find:

$$V_{TS} = \frac{\rho_p d^2 g C_c}{18\eta}$$

where

V_{TS} is the terminal settling velocity of the particle.

The terminal velocity can also be derived for other kinds of forces. If Stokes' law holds, then the resistance to motion is directly proportional to speed. The constant of proportionality is the mechanical mobility (*B*) of a particle:

$$B = \frac{V}{F_D} = \frac{C_c}{3\pi\eta d}$$

A particle travelling at any reasonable initial velocity approaches its terminal velocity exponentially with an *e*-folding time equal to the relaxation time:

$$V(t) = V_f - (V_f - V_0)e^{-\frac{t}{\tau}}$$

where:

$V(t)$ is the particle speed at time t

V_f is the final particle speed

V_0 is the initial particle speed

To account for the effect of the shape of non-spherical particles, a correction factor known as the *dynamic shape factor* is applied to Stokes' law. It is defined as the ratio of the resistive force of the irregular particle to that of a spherical particle with the same volume and velocity:

$$\chi = \frac{F_D}{3\pi\eta V d_e}$$

where:

X is the dynamic shape factor

Aerodynamic Diametre

The aerodynamic diametre of an irregular particle is defined as the diametre of the spherical particle with a density of 1000 kg/m^3 and the same settling velocity as the irregular particle.

Neglecting the slip correction, the particle settles at the terminal velocity proportional to the square of the aerodynamic diametre, d_a:

$$V_{TS} = \frac{\rho_0 d_a^2 g}{18\eta}$$

where

ρ_0 = standard particle density (1000 kg/m^3).

This equation gives the aerodynamic diametre:

$$d_a = d_e \left(\frac{\rho_p}{\rho_0 \chi} \right)^{\frac{1}{2}}$$

One can apply the aerodynamic diametre to particulate pollutants or to inhaled drugs to predict where in the respiratory tract such particles deposit. Pharmaceutical companies typically use aerodynamic diametre, not geometric diametre, to characterize particles in inhalable drugs.

Dynamics

The previous discussion focussed on single aerosol particles. In contrast, *aerosol dynamics* explains the evolution of complete aerosol populations. The concentrations of particles will change over time as a result of many processes. External processes that move particles outside a volume of gas under study include diffusion, gravitational settling, and electric charges and other external forces that cause particle migration. A second set of processes internal to a given volume of gas include particle formation (nucleation), evaporation, chemical reaction, and coagulation.

A differential equation called the *Aerosol General Dynamic Equation* (GDE) characterizes the evolution of the number density of particles in an aerosol due to these processes.

$$\frac{\partial n_i}{\partial t} = -\nabla \cdot n_i \mathrm{q} + \nabla \cdot D_p \nabla_i + \left(\frac{\partial n_i}{\partial t} \right)_{growth} + \left(\frac{\partial n_i}{\partial t} \right)_{coag} - \nabla \cdot \mathrm{q}_F n_i$$

Change in time = Convective transport + brownian diffusion + gas-particle interactions + coagulation + migration by external forces

Where:

n_i is number density of particles of size category i

q is the particle velocity

D_p is the particle Stokes-Einstein diffusivity

q_F is the particle velocity associated with an external force

Coagulation

As particles and droplets in an aerosol collide with one another, they may undergo coalescence or aggregation. This process leads to a change in the aerosol particle-size distribution, with the mode increasing in diametre as total number of particles decreases. On occasion, particles may shatter apart into numerous smaller particles; however, this process usually occurs primarily in particles too large for consideration as aerosols.

Dynamics Regimes

The Knudsen number of the particle define three different dynamical regimes that govern the behaviour of an aerosol:

$$K_n = \frac{2\lambda}{d}$$

where λ is the mean free path of the suspending gas and d is the diametre of the particle. For particles in the *free molecular regime*, $K_n >> 1$; particles small compared to the mean free path of the suspending gas. In this regime, particles interact with the suspending gas through a series of "ballistic" collisions with gas molecules. As such, they behave similarly to gas molecules, tending to follow streamlines and diffusing rapidly through Brownian motion. The mass flux equation in the free molecular regime is:

$$I = \frac{\pi a^2}{k_b}\left(\frac{P_\infty}{T_\infty} - \frac{P_A}{T_A}\right)\cdot C_A \alpha$$

where a is the particle radius, $P\_$ and P_A are the pressures far from the droplet and at the surface of the droplet respectively, k_b is the Boltzmann constant, T is the temperature, C_A is mean thermal velocity and α is mass accommodation coefficient. The derivation of this equation assumes constant pressure and constant diffusion coefficient.

Particles are in the *continuum regime* when $K_n << 1$. In this regime, the particles are big compared to the mean free path of the suspending gas, meaning that the suspending gas acts as a continuous fluid flowing round the particle. The molecular flux in this regime is:

$$I_{cont} \sim \frac{4\pi a M_A D_{AB}}{RT}\left(P_{A\infty} - P_{AS}\right)$$

where a is the radius of the particle A, M_A is the molecular mass of the particle A, D_{AB} is the diffusion coefficient between particles A and B, R is the ideal gas constant, T is the temperature (in absolute units like kelvin), and P_{A-} and P_{AS} are the pressures at infinite and at the surface respectively.

The *transition regime* contains all the particles in between the free molecular and continuum regimes or $K_n \approx 1$. The forces experienced by a particle are a complex combination of interactions with individual gas molecules and macroscopic interactions. The semi-empirical equation describing mass flux is:

$$I = I_{cont} \cdot \frac{1 + K_n}{1 + 1.71K_n + 1.33K_n^2}$$

where I_{cont} is the mass flux in the continuum regime. This formula is called the Fuchs-Sutugin interpolation formula. These equations do not take into account the heat release effect.

Partitioning

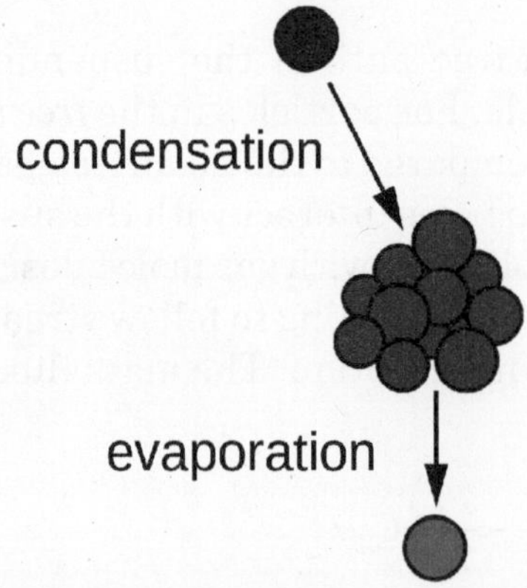

Figure: *Condensation and evaporation*

Aerosol partitioning theory governs condensation on and evaporation from an aerosol surface, respectively. Condensation of mass causes the mode of the particle-size distributions of the aerosol to increase; conversely, evaporation causes the mode to decrease. Nucleation is the process of forming aerosol mass from the condensation of a gaseous precursor, specifically a vapour. Net condensation of the vapour requires supersaturation, a partial pressure greater than its vapour pressure. This can happen for three reasons:

1. Lowering the temperature of the system lowers the vapour pressure.
2. Chemical reactions may increase the partial pressure of a gas or lower its vapour pressure.
3. The addition of additional vapour to the system may lower the equilibrium vapour pressure according to Raoult's law.

There are two types of nucleation processes. Gases preferentially condense onto surfaces of pre-existing aerosol particles, known as heterogeneous nucleation. This process causes the diametre at the mode of particle-size distribution to increase with constant number concentration. With sufficiently high supersaturation and no suitable surfaces, particles may condense in the absence of a pre-existing surface, known as homogeneous nucleation. This results in the addition of very small, rapidly growing particles to the particle-size distribution.

Activation

Water coats particles in an aerosols, making them *activated*, usually in the context of forming a cloud droplet. Following the Kelvin equation (based on the curvature of liquid droplets), smaller particles need a higher ambient relative humidity to maintain equilibrium than larger particles do. The following formula gives relative humidity at equilibrium:

$$RH = \frac{p_s}{p_0} \times 100\% = S \times 100\%$$

where p_s is the saturation vapour pressure above a particle at equilibrium (around a curved liquid droplet), p_0 is the saturation vapour pressure (flat surface of the same liquid) and S is the saturation ratio.

Kelvin equation for saturation vapour pressure above a curved surface is:

$$\ln \frac{p_s}{p_0} = \frac{2\sigma M}{RT\rho \cdot r_p}$$

where r_p droplet radius, σ surface tension of droplet, ρ density of liquid, M molar mass, T temperature, and R molar gas constant.

Solution to the General Dynamic Equation

There are no general solutions to the general dynamic equation (GDE); common methods used to solve the general dynamic equation include:

- Moment method,
- Modal/sectional method, and
- Quadrature method of moments.

Generation and Applications

People generate aerosols for various purposes, including:

- as test aerosols for calibrating instruments, performing research, and testing sampling equipment and air filters;
- to deliver deodorants, paints, and other consumer products in sprays;

- for dispersal and agricultural application of pesticides;
- for medical treatment of respiratory disease; and
- in fuel injection systems and other combustion technology.

Some devices for generating aerosols are:

- Aerosol spray
- Atomizer nozzle or Nebulizer
- Electrospray
- Vibrating orifice aerosol generator (VOAG)

Detection

Aerosol can either be measured in-situ or with remote sensing techniques.

In Situ Observations

Some available in situ measurement techniques include:

- Aerosol mass spectrometre (AMS)
- Differential mobility analyzer (DMA)
- Aerodynamic particle sizer (APS)
- Wide range particle spectrometre (WPS)
- Micro-Orifice Uniform Deposit Impactor(MOUDI)
- Condensation particle counter (CPC)
- Epiphaniometre
- Electrical low pressure impactor (ELPI)
- Aerosol particle mass-analyser (APM)

Remote Sensing Approach

Remote sensing approaches include:

- Sun photometre
- Lidar

Size Selective Sampling

Particles can deposit in the nose, mouth, pharynx and larynx (the head airways region), deeper within the respiratory tract (from the trachea to the terminal bronchioles), or in the alveolar region. The location of deposition of aerosol particles within the in the respiratory system strongly determines the health effects of exposure to such aerosols. This phenomenon led people to invent aerosol samplers that select a subset of the aerosol particles that reach certain parts of the respiratory system. Examples of these subsets of the particle-size distribution of an aerosol, important in occupational health, include the inhalable, thoracic, and respirable fractions. The fraction that can enter each part of the respiratory system depends on the deposition of particles in the upper parts of the airway. The inhalable fraction of particles, defined as the proportion of particles originally in the air that can enter the nose or mouth, depends on external wind speed and direction and on the particle-size distribution by aerodynamic diametre. The thoracic fraction is the proportion of the particles in ambient aerosol that can reach the thorax or chest region. The respirable fraction is the proportion of particles in the air that can reach the alveolar region. To measure the respirable fraction of particles in air, a pre-collector is used with a sampling filter. The pre-collector excludes particles as the airways remove particles from inhaled air. The sampling filter collects the particles for measurement. It is common to use cyclonic separation for the pre-collector, but other techniques include impactors, horizontal elutriators, and large pore membrane filters.

Two alternative size selective criteria, often used in atmospheric monitoring are PM10 and PM2.5. PM10 is defined by ISO as *particles which pass through a size-selective inlet with a 50 % efficiency cut-off at 10 μm aerodynamic diametre. PM10 corresponds to the "thoracic convention" as defined in ISO 7708:1995, Clause 6* and PM2.5 as *particles which pass through a size-selective inlet with a 50 % efficiency cut-off at 2,5 μm aerodynamic diametre. PM2,5 corresponds to the "high-risk respirable convention" as defined in ISO 7708:1995, 7.1.* The United States Environmental Protection Agency replaced the older standards for particulate matter based on Total Suspended Particulate with another standard based on PM10 in 1987 and then introduced standards for PM2.5 (also known as fine particulate matter) in 1997.

Atmospheric

Although all hydrometeors, solid and liquid, can be described as aerosols, a distinction is commonly made between such dispersions (i.e. clouds) containing activated drops and crystals, and aerosol particles. Atmosphere of Earth contains aerosols of various types and concentrations, including quantities of:

- natural inorganic materials: fine dust, sea salt, water droplets.
- natural organic materials: smoke, pollen, spores, bacteria
- anthropogenic products of combustion such as: smoke, ashes or dusts

Aerosols can be found in urban Ecosystems in various forms, for example:

- Dust,
- Cigarette smoke,
- Mist from aerosol spray cans,
- Soot or fumes in car exhaust.

The presence of aerosols in earth's atmosphere can influence Earth's climate, as well as human health.

Importance of Aerosols

- Volcanic eruptions release large amounts of sulphuric acid, hydrogen sulphide and hydrochloric acid into the atmosphere. These gases represent aerosols and eventually return to earth as acid rain, having a number of adverse effects on the environment and human life.
- Aerosols interact with the Earth's energetic budget in two ways, directly and indirectly.

  E.g., a *direct* effect is that aerosols scatter sunlight directly back into space. This can lead to a significant decrease in the temperature, being an additional element to the greenhouse effect and therefore contributing to the global climate change.

The *indirect* effects refer to the aerosols interfering with formations that interact directly with radiation. For example, they are able to modify the size of the cloud particles in the lower atmosphere, thereby changing the way clouds reflect and absorb light and therefore modifying the Earth's energy budget.

- When aerosols absorb or adsorb pollutants, it facilitates the deposition of pollutants to the surface of the earth as well as to bodies of water. This has the potential to be damaging to both the environment and human health.
- Aerosol particles with an effective diametre smaller than 10 μm can enter the bronchi, while the ones with an effective diametre smaller than 2.5 μm can enter as far as the gas exchange region in the lungs, which can be hazardous to human health.

Chapter 4

Non Variable and Variable Components

Non-variable components

Some gases of the atmosphere remain constant at surface of globe up to the height of 80 to 88 km. This is due to transportation of gases on continental level, diffusion of gases, turbulent mixing and convection. These gases are called non-variable components. They are Non-variable components (permanent constituents)

Nitrogen

Nitrogen is a chemical element with symbol N and atomic number 7. It is the lightest pnictogen and at room temperature, it is a colourless and odourless diatomic gas. Nitrogen is a common element in the universe, estimated at about seventh in total abundance in our galaxy and the Solar System. On Earth, the element forms about 78% of Earth's atmosphere and as such is the most abundant pure element. The element nitrogen was discovered as a separable component of air, by Scottish physician Daniel Rutherford, in 1772.

Many industrially important compounds, such as ammonia, nitric acid, organic nitrates (propellants and explosives), and cyanides, contain nitrogen. The extremely strong triple bond in elemental nitrogen ($N \equiv N$) dominates nitrogen chemistry, causing difficulty for both organisms and industry in converting the N_2 into useful compounds, but at the same time causing release of large amounts of often useful energy when the compounds burn, explode, or decay back into nitrogen gas. Synthetically-produced ammonia and nitrates are key industrial fertilizers and fertilizer nitrates are key pollutants in causing the eutrophication of water systems. Outside their major uses as fertilizers and energy-stores, nitrogen compounds are versatile organics. Nitrogen is part of materials as diverse as Kevlar fabric and cyanoacrylate "super" glue. Nitrogen is a constituent of molecules in every major pharmacological drug class, including the antibiotics.

Many drugs are mimics or prodrugs of natural nitrogen-containing signal molecules: for example, the organic nitrates nitroglycerin and nitroprusside control blood pressure by being metabolized to natural nitric oxide. Plant alkaloids (often defence chemicals) contain nitrogen by definition, and thus many notable nitrogen-containing drugs, such as caffeine and morphine are either alkaloids or synthetic mimics that act (as many plant alkaloids do) upon receptors of animal neurotransmitters (for example, synthetic amphetamines).

Nitrogen occurs in all organisms, primarily in amino acids (and thus proteins) and also in the nucleic acids (DNA and RNA). The human body contains about 3% by mass of nitrogen, the fourth most abundant element in the body after oxygen, carbon, and hydrogen. The nitrogen cycle describes movement of the element from the air, into the biosphere and organic compounds, then back into the atmosphere.

Oxygen

Oxygen is a chemical element with symbol O and atomic number 8. It is a member of the chalcogen group on the periodic table and is a highly reactive nonmetallic element and oxidizing agent that readily forms compounds (notably oxides) with most elements.

By mass, oxygen is the third-most abundant element in the universe, after hydrogen and helium. At STP, two atoms of the element bind to form dioxygen, a diatomic gas that is colourless, odourless, and tasteless, with the formula O_2.

Many major classes of organic molecules in living organisms, such as proteins, nucleic acids, carbohydrates, and fats, contain oxygen, as do the major inorganic compounds that are constituents of animal shells, teeth, and bone. Most of the mass of living organisms is oxygen as it is a part of water, the major constituent of lifeforms (for example, about two-thirds of human body mass).

Elemental oxygen is produced by cyanobacteria, algae and plants, and is used in cellular respiration for all complex life. Oxygen is toxic to obligately anaerobic organisms, which were the dominant form of early life on Earth until O_2 began to accumulate in the atmosphere. Free elemental O_2 only began to accumulate in the atmosphere about

2.5 billion years ago during the Great Oxygenation Event, about a billion years after the first appearance of these organisms. Diatomic oxygen gas constitutes 20.8% of the volume of air. Oxygen is the most abundant element by mass in the Earth's crust as part of oxide compounds such as silicon dioxide, making up almost half of the crust's mass.

Oxygen is an important part of the atmosphere, and is necessary to sustain most terrestrial life as it is used in respiration. However, it is too chemically reactive to remain a free element in Earth's atmosphere without being continuously replenished by the photosynthetic action of living organisms, which use the energy of sunlight to produce elemental oxygen from water.

Another form (allotrope) of oxygen, ozone (O_3), strongly absorbs UVB radiation and consequently the high-altitude ozone layer helps protect the biosphere from ultraviolet radiation, but is a pollutant near the surface where it is a by-product of smog. At even higher low earth orbit altitudes, atomic oxygen is a significant presence and a cause of erosion for spacecraft. Oxygen is produced industrially by fractional distillation of liquefied air, use of zeolites with pressure-cycling to concentrate oxygen from air, electrolysis of water and other means. Uses of elemental oxygen include the production of steel, plastics and textiles, brazing, welding and cutting of steels and other metals, rocket propellant, oxygen therapy and life support systems in aircraft, submarines, spaceflight and diving.

Oxygen was discovered independently by Carl Wilhelm Scheele, in Uppsala, in 1773 or earlier, and Joseph Priestley in Wiltshire, in 1774, but Priestley is often given priority because his work was published first. The name *oxygen* was coined in 1777 by Antoine Lavoisier, whose experiments with oxygen helped to discredit the then-popular phlogiston theory of combustion and corrosion. Its name derives from the Greek roots *oxys*, "acid", literally "sharp", referring to the sour taste of acids and *genes*, "producer", literally "begetter", because at the time of naming, it was mistakenly thought that all acids required oxygen in their composition.

Argon

Argon is a chemical element with symbol Ar and atomic number 18. It is in group 18 of the periodic table and is a noble gas. Argon is the third most common gas in the Earth's atmosphere, at 0.93% (9,300 ppm), making it approximately 23.8 times as abundant as the next most common atmospheric gas, carbon dioxide (390 ppm), and more than 500 times as abundant as the next most common noble gas, neon

(18 ppm). Nearly all of this argon is radiogenic argon-40 derived from the decay of potassium-40 in the Earth's crust. In the universe, argon-36 is by far the most common argon isotope, being the preferred argon isotope produced by stellar nucleosynthesis in supernovas.

The name "argon" is derived from the Greek word *αρãïí*, neuter singular form of *αρãïò* meaning "lazy" or "inactive", as a reference to the fact that the element undergoes almost no chemical reactions. The complete octet (eight electrons) in the outer atomic shell makes argon stable and resistant to bonding with other elements. Its triple point temperature of 83.8058 K is a defining fixed point in the International Temperature Scale of 1990.

Argon is produced industrially by the fractional distillation of liquid air. Argon is mostly used as an inert shielding gas in welding and other high-temperature industrial processes where ordinarily non-reactive substances become reactive; for example, an argon atmosphere is used in graphite electric furnaces to prevent the graphite from burning. Argon gas also has uses in incandescent and fluorescent lighting, and other types of gas discharge tubes. Argon makes a distinctive blue-green gas laser.

Carbon Dioxide

Carbon dioxide (chemical formula CO_2) is a naturally occurring chemical compound composed of 2 oxygen atoms each covalently double bonded to a single carbon atom. It is a gas at standard temperature and pressure and exists in Earth's atmosphere in this state, as a trace gas at a concentration of 0.04 per cent (400 ppm) by volume, as of 2014.

As part of the carbon cycle, plants, algae, and cyanobacteria use light energy to photosynthesize carbohydrate from carbon dioxide and water, with oxygen produced as a waste product. However, photosynthesis cannot occur in darkness and at night some carbon dioxide is produced by plants during respiration. It is produced during the respiration of all other aerobic organisms and is exhaled in the breath of air-breathing land animals, including humans. Carbon dioxide is produced during the processes of decay of organic materials and the fermentation of sugars in beer and winemaking. It is produced by combustion of wood, carbohydrates and major carbon- and hydrocarbon-rich fossil fuels such as coal, peat, petroleum and natural gas. It is emitted from volcanoes, hot springs and geysers and is freed from carbonate rocks by dissolution in water and acids. CO_2 is found in lakes, at depth under the sea and commingled with oil and gas deposits.

The environmental effects of carbon dioxide are of significant interest. Atmospheric carbon dioxide is the primary source of carbon in life on Earth and its concentration in Earth's pre-industrial atmosphere since late in the Precambrian eon was regulated by photosynthetic organisms. Carbon dioxide is an important greenhouse gas and burning of carbon-based fuels since the industrial revolution has rapidly increased its concentration in the atmosphere, leading to global warming. It is also a major source of ocean acidification since it dissolves in water to form carbonic acid.

Neon

Neon is a chemical element with symbol Ne and atomic number 10. It is in group 18 (noble gases) of the periodic table. Neon is a colourless, odourless, inert monatomic gas under standard conditions, with about two-thirds the density of air. It was discovered (along with krypton and xenon) in 1898 as one of the three residual rare inert elements remaining in dry air, after nitrogen, oxygen, argon and carbon dioxide are removed. Neon was the second of these three rare gases to be discovered, and was immediately recognized as a new element from its bright red emission spectrum. Neon is chemically inert and forms no uncharged chemical compounds.

During cosmic nucleogenesis of the elements, large amounts of neon are built up from the alpha-capture fusion process in stars. Although neon is a very common element in the universe and solar system (it is fifth in cosmic abundance after hydrogen, helium, oxygen and carbon), it is very rare on Earth. It composes about 18.2 ppm of air by volume (this is about the same as the molecular or mole fraction), and a smaller fraction in Earth's crust. The reason for neon's relative scarcity on Earth and the inner (terrestrial) planets, is that neon forms no compounds to fix it to solids, and is highly volatile, therefore escaping from the planetesimals under the warmth of the newly ignited Sun in the early Solar System. Even the atmosphere of Jupiter is somewhat depleted of neon, presumably for this reason.

Neon gives a distinct reddish-orange glow when used in either low-voltage neon glow lamps or in high-voltage discharge tubes or neon advertising signs. The red emission line from neon is also responsible for the well known red light of helium–neon lasers. Neon is used in a few plasma tube and refrigerant applications but has few other commercial uses. It is commercially extracted by the fractional distillation of liquid air. It is considerably more expensive than helium, since air is its only source.

Helium

Helium is a chemical element with symbol He and atomic number 2. It is a colourless, odourless, tasteless, non-toxic, inert, monatomic gas that heads the noble gas group in the periodic table. Its boiling and melting points are the lowest among the elements and it exists only as a gas except in extreme conditions.

Helium is the second lightest element and is the second most abundant element in the observable universe, being present at about 24% of the total elemental mass, which is more than 12 times the mass of all the heavier elements combined. Its abundance is similar to this figure in the Sun and in Jupiter. This is due to the very high nuclear binding energy (per nucleon) of helium-4 with respect to the next three elements after helium. This helium-4 binding energy also accounts for why it is a product of both nuclear fusion and radioactive decay. Most helium in the universe is helium-4, and is believed to have been formed during the Big Bang. Large amounts of new helium are being created by nuclear fusion of hydrogen in stars.

Helium is named for the Greek god of the Sun, Helios. It was first detected as an unknown yellow spectral line signature in sunlight during a solar eclipse in 1868 by French astronomer Jules Janssen. Janssen is jointly credited with detecting the element along with Norman Lockyer. Jannsen observed during the solar eclipse of 1868 while Lockyer observed from Britain. Lockyer was the first to propose that the line was due to a new element, which he named.

The formal discovery of the element was made in 1895 by two Swedish chemists, Per Teodor Cleve and Nils Abraham Langlet, who found helium emanating from the uranium ore cleveite. In 1903, large reserves of helium were found in natural gas fields in parts of the United States, which is by far the largest supplier of the gas today.

Helium is used in cryogenics (its largest single use, absorbing about a quarter of production), particularly in the cooling of superconducting magnets, with the main commercial application being in MRI scanners. Helium's other industrial uses—as a pressurizing and purge gas, as a protective atmosphere for arc welding and in processes such as growing crystals to make silicon wafers—account for half of the gas produced.

A well-known but minor use is as a lifting gas in balloons and airships. As with any gas whose density differs from that of air, inhaling a small volume of helium temporarily changes the timbre and quality of the human voice.

In scientific research, the behaviour of the two fluid phases of helium-4 (helium I and helium II) is important to researchers studying quantum mechanics (in particular the property of superfluidity) and to those looking at the phenomena, such as superconductivity, produced in matter near absolute zero.

On Earth it is relatively rare — 5.2 ppm by volume in the atmosphere. Most terrestrial helium present today is created by the natural radioactive decay of heavy radioactive elements (thorium and uranium, although there are other examples), as the alpha particles emitted by such decays consist of helium-4 nuclei. This radiogenic helium is trapped with natural gas in concentrations up to 7% by volume, from which it is extracted commercially by a low-temperature separation process called fractional distillation. Helium is a finite resource and is one of the few elements with escape velocity, meaning that once released into the atmosphere, it escapes into space.

Xenon

Xenon is a chemical element with symbol Xe and atomic number 54. It is a colourless, dense, odourless noble gas, that occurs in the Earth's atmosphere in trace amounts. Although generally unreactive, xenon can undergo a few chemical reactions such as the formation of xenon hexafluoroplatinate, the first noble gas compound to be synthesized.

Naturally occurring xenon consists of eight stable isotopes. There are also over 40 unstable isotopes that undergo radioactive decay. The isotope ratios of xenon are an important tool for studying the early history of the Solar System. Radioactive xenon-135 is produced from iodine-135 as a result of nuclear fission, and it acts as the most significant neutron absorber in nuclear reactors.

Xenon is used in flash lamps and arc lamps, and as a general anesthetic. The first excimer laser design used a xenon dimer molecule (Xe_2) as its lasing medium, and the earliest laser designs used xenon flash lamps as pumps. Xenon is also being used to search for hypothetical weakly interacting massive particles and as the propellant for ion thrusters in spacecraft.

Hydrogen

Hydrogen is a chemical element with chemical symbol H and atomic number 1. With an atomic weight of 1.00794 u, hydrogen is the lightest element on the periodic table. Its monatomic form (H) is the most abundant chemical substance in the universe, constituting roughly

75% of all baryonic mass. Non-remnant stars are mainly composed of hydrogen in its plasma state. The most common isotope of hydrogen, termed *protium* (name rarely used, symbol ^1H), has a single proton and zero neutrons.

The universal emergence of atomic hydrogen first occurred during the recombination epoch. At standard temperature and pressure, hydrogen is a colourless, odourless, tasteless, non-toxic, nonmetallic, highly combustible diatomic gas with the molecular formula H_2. Since hydrogen readily forms covalent compounds with most non-metallic elements, most of the hydrogen on Earth exists in molecular forms such as in the form of water or organic compounds.

Hydrogen plays a particularly important role in acid–base reactions as many acid-base reactions involve the exchange of protons between soluble molecules. In ionic compounds, hydrogen can take the form of a negative charge (i.e., anion) known as a hydride, or as a positively charged (i.e., cation) species denoted by the symbol H^+. The hydrogen cation is written as though composed of a bare proton, but in reality, hydrogen cations in ionic compounds are always more complex species than that would suggest.

As the only neutral atom for which the Schrödinger equation can be solved analytically, study of the energetics and bonding of the hydrogen atom has played a key role in the development of quantum mechanics.

Hydrogen gas was first artificially produced in the early 16th century, via the mixing of metals with acids. In 1766–81, Henry Cavendish was the first to recognize that hydrogen gas was a discrete substance, and that it produces water when burned, a property which later gave it its name: in Greek, hydrogen means "water-former".

Industrial production is mainly from the steam reforming of natural gas, and less often from more energy-intensive hydrogen production methods like the electrolysis of water. Most hydrogen is employed near its production site, with the two largest uses being fossil fuel processing (e.g., hydrocracking) and ammonia production, mostly for the fertilizer market.

Hydrogen is a concern in metallurgy as it can embrittle many metals, complicating the design of pipelines and storage tanks.

Methane

Methane is a chemical compound with the chemical formula CH_4 (one atom of carbon and four atoms of hydrogen). It is the simplest alkane and the main component of natural gas. The relative abundance of methane makes it an attractive fuel, though capturing and storing it poses challenges due to its gaseous state found at normal conditions. In its natural state, methane is found both below ground, and under the sea floor, where it often finds its way to the surface and in the earth's atmosphere where it is known as atmospheric methane.

Methane is a tetrahedral molecule with four equivalent C-H bonds. Its electronic structure is described by four bonding molecular orbitals (MOs) resulting from the overlap of the valence orbitals on C and H. The lowest energy MO is the result of the overlap of the 2s orbital on carbon with the in-phase combination of the 1s orbitals on the four hydrogen atoms. Above this level in energy is a triply degenerate set of MOs that involve overlap of the 2p orbitals on carbon with various linear combinations of the 1s orbitals on hydrogen. The resulting "three-over-one" bonding scheme is consistent with photoelectron spectroscopic measurements.

At room temperature and standard pressure, methane is a colourless, odourless gas. The familiar smell of natural gas as used in homes is a safety measure achieved by the addition of an odorant, usually blends containing tert-butylthiol. Methane has a boiling point of –161 °C (–257.8 °F) at a pressure of one atmosphere. As a gas it is flammable over a range of concentrations (4.4–17%) in air at standard pressure.

Nitrous Oxide

Nitrous oxide, commonly known as laughing gas, nitrous, nitro, or NOS is a chemical compound with the formula N_2O. It is an oxide of nitrogen. At room temperature, it is a colourless, non-flammable gas, with a slightly sweet odour and taste. It is used in surgery and dentistry for its anaesthetic and analgesic effects. It is known as "laughing gas" due to the euphoric effects of inhaling it, a property that has led to its recreational use as a dissociative anaesthetic. It is also used as an oxidizer in the launching of rockets and in motor racing to increase the power output of engines. At elevated temperatures, nitrous oxide is a powerful oxidizer similar to molecular oxygen.

Nitrous oxide gives rise to nitric oxide (NO) on reaction with oxygen atoms, and this NO in turn reacts with ozone. As a result, it is the main naturally occurring regulator of stratospheric ozone. It is also a major greenhouse gas and air pollutant. Considered over a 100-year period, it has 298 times more impact per unit mass (global warming potential) than carbon dioxide.

It is on the World Health Organization's List of Essential Medicines, the most important medications needed in a health system.

The gas is approved for use as a food additive (also known as E942), specifically as an aerosol spray propellant. Its most common uses in this context are in aerosol whipped cream canisters, cooking sprays, and as an inert gas used to displace oxygen, to inhibit bacterial growth, when filling packages of potato chips and other similar snack foods.

The gas is extremely soluble in fatty compounds. In aerosol whipped cream, it is dissolved in the fatty cream until it leaves the can, when it becomes gaseous and thus creates foam. Used in this way, it produces whipped cream four times the volume of the liquid, whereas whipping air into cream only produces twice the volume.

If air were used as a propellant, oxygen would accelerate rancidification of the butterfat; nitrous oxide inhibits such degradation. Carbon dioxide cannot be used for whipped cream because it is acidic in water, which would curdle the cream and give it a seltzer-like "sparkling" sensation.

However, the whipped cream produced with nitrous oxide is unstable and will return to a more liquid state within half an hour to one hour. Thus, the method is not suitable for decorating food that will not be immediately served.

Similarly, cooking spray, which is made from various types of oils combined with lecithin (an emulsifier), may use nitrous oxide as a propellant; other propellants used in cooking spray include food-grade alcohol and propane.

Users of nitrous oxide often obtain it from whipped cream dispensers that use nitrous oxide as a propellant, for recreational use as a euphoria-inducing inhalant drug. It is not harmful in small doses, but risks due to lack of oxygen do exist.

Radon

Radon is a chemical element with symbol Rn and atomic number 86. It is a radioactive, colourless, odourless, tasteless noble gas, occurring naturally as an indirect decay product of uranium or thorium. Its most stable isotope, ^{222}Rn, has a half-life of 3.8 days. Radon is one of the densest substances that remains a gas under normal conditions. It is also the only gas under normal conditions that only has radioactive isotopes, and is considered a health hazard due to its radioactivity. Intense radioactivity has also hindered chemical studies of radon and only a few compounds are known.

Radon is formed as one intermediate step in the normal radioactive decay chains, through which thorium and uranium slowly decay into lead. Thorium and uranium are the two most common radioactive elements on earth; they have been around since the earth was formed. Their naturally occurring isotopes have very long half-lives, on the order of billions of years.

Thorium and uranium, their decay product radium, and its decay product radon, will therefore continue to occur for tens of millions of years at almost the same concentrations as they do now. As radon itself decays, it produces new radioactive elements called radon daughters or decay products.

Unlike the gaseous radon itself, radon daughters are solids and stick to surfaces, such as dust particles in the air. If such contaminated dust is inhaled, these particles can stick to the airways of the lung and increase the risk of developing lung cancer.

Unlike all the other intermediate elements in the aforementioned decay chains, radon is gaseous and easily inhaled. Thus, even in this age of nuclear reactors, naturally-occurring radon is responsible for the majority of the public exposure to ionizing radiation. It is often the single largest contributor to an individual's background radiation dose, and is the most variable from location to location.

Despite its short lifetime, some radon gas from natural sources can accumulate to far higher than normal concentrations in buildings, especially in low areas such as basements and crawl spaces due to its heavy nature. It can also be found in some spring waters and hot springs.

Epidemiological studies have shown a clear link between breathing high concentrations of radon and incidence of lung cancer. Thus, radon is considered a significant contaminant that affects indoor air quality worldwide. According to the United States Environmental Protection Agency, radon is the second most frequent cause of lung cancer, after cigarette smoking, causing 21,000 lung cancer deaths per year in the United States. About 2,900 of these deaths occur among people who have never smoked. While radon is the second most frequent cause of lung cancer, it is the number one cause among non-smokers, according to EPA estimates.

Variable Components

Some gases or components of the atmosphere ch anges with change with change in time, palce, season etc., and these components are called as variable component they are-

Water Vapour

Water vapour, or water vapour or aqueous vapour, is the gaseous phase of water. It is one state of water within the hydrosphere. Water vapour can be produced from the evaporation or boiling of liquid water or from the sublimation of ice.

Unlike other forms of water, water vapour is invisible. Under typical atmospheric conditions, water vapour is continuously generated by evaporation and removed by condensation. It is lighter than air and triggers convection currents that can lead to clouds.

Water vapour is a relatively common atmospheric constituent, present even in the solar atmosphere as well as every planet in the Solar System and many astronomical objects including natural satellites, comets and even large asteroids. Likewise the detection of extrasolar water vapour would indicate a similar distribution in other planetary systems. Water vapour is significant in that it can be indirect evidence supporting the presence of extraterrestrial liquid water in the case of some planetary mass objects.

Being a component of Earth's hydrosphere and hydrologic cycle, it is particularly abundant in Earth's atmosphere where it is also a potent greenhouse gas along with other gases such as carbon dioxide and methane. Use of water vapour, as steam, has been important to humans for cooking and as a major component in energy production and transport systems since the industrial revolution.

Occurrence

Water vapour is a significant component of the Earth's atmosphere and a greenhouse gas. It is also common in the Solar System and by extension, other planetary systems. Its signature has been detected in the atmospheres of the Sun, occurring in sunspots. The presence of water vapour has been detected in the atmospheres of Mercury, Venus, Earth (and Moon), Mars, Jupiter, Saturn, Uranus and Neptune, the planets of the Solar System, although typically in only trace amounts.

Water vapour ejected into the atmosphere of some planetary mass objects (such as water ejected from geological formations such as geysers) may indicate the presence of substantial quantities of subsurface water. Plumes of water vapour have been detected on Europa (a moon of Jupiter) and are similar to plumes of water vapour detected on Enceladus (a moon of Saturn).

Traces of water vapour have also been detected in the stratosphere of Titan. Water vapour has been found to be a major constituent of the atmosphere of dwarf planet, Ceres, largest object in the asteroid belt The detection was made by using the far-infrared abilities of the Herschel Space Observatory.

The finding is unexpected because comets, not asteroids, are typically considered to "sprout jets and plumes." According to one of the scientists, "The lines are becoming more and more blurred between comets and asteroids." Scientists studying Mars hypothesize that if water moves about the planet, it does so as vapour.

The brilliance of comet tails comes largely from water vapour. On approach to the sun, the ice many comets carry sublimates to vapour, which reflects light from the sun. Knowing a comet's distance from the sun, astronomers may deduce a comet's water content from its brilliance.

Water vapour has also been confirmed outside the Solar System. Spectroscopic analysis of HD 209458 b, an extrasolar planet in the constellation Pegasus, provides the first evidence of atmospheric water vapour beyond the Solar System.

A star called CW Leonis was found to have a ring of vast quantities of water vapour circling the aging, massive star. A NASA satellite designed to study chemicals in interstellar gas clouds, made the discovery with an onboard spectrometre.

Most likely, "the water vapour was vaporized from the surfaces of orbiting comets." HAT-P-11b a relatively small exoplanet has also been found to possess water vapour.

General Properties of Water Vaporization

Evaporation and Sublimation

Whenever a water molecule leaves a surface and diffuses into a surrounding gas, it is said to have evaporated. Each individual water molecule which transitions between a more associated (liquid) and a less associated (vapour/gas) state does so through the absorption or release of kinetic energy. The aggregate measurement of this kinetic energy transfer is defined as thermal energy and occurs only when there is differential in the temperature of the water molecules. Liquid water that becomes water vapour takes a parcel of heat with it, in a process called evaporative cooling. The amount of water vapour in the air determines how fast each molecule will return to the surface. When a net evaporation occurs, the body of water will undergo a net cooling directly related to the loss of water.

In the US, the National Weather Service measures the actual rate of evaporation from a standardized "pan" open water surface outdoors, at various locations nationwide. Others do likewise around the world. The US data is collected and compiled into an annual evaporation map. The measurements range from under 30 to over 120 inches per year.

Formulas can be used for calculating the rate of evaporation from a water surface such as a swimming pool. In some countries, the evaporation rate far exceeds the precipitation rate.

Evaporative cooling is restricted by atmospheric conditions. Humidity is the amount of water vapour in the air. The vapour content of air is measured with devices known as hygrometres. The measurements are usually expressed as specific humidity or percent relative humidity.

The temperatures of the atmosphere and the water surface determine the equilibrium vapour pressure; 100% relative humidity occurs when the partial pressure of water vapour is equal to the equilibrium vapour pressure.

This condition is often referred to as complete saturation. Humidity ranges from 0 gram per cubic metre in dry air to 30 grams per cubic metre (0.03 ounce per cubic foot) when the vapour is saturated at 30 °C.

Another form of evaporation is sublimation, by which water molecules become gaseous directly, leaving the surface of ice without first becoming liquid water. Sublimation accounts for the slow mid-winter disappearance of ice and snow at temperatures too low to cause melting. Antarctica shows this effect to a unique degree because it is by far the continent with the lowest rate of precipitation on Earth.

As a result there are large areas where millennial layers of snow have sublimed, leaving behind whatever non-volatile materials they had contained. This is extremely valuable to certain scientific disciplines, a dramatic example being the collection of meteorites that are left exposed in unparalleled numbers and excellent states of preservation.

Sublimation is of importance in the preparation of certain classes of biological specimens for scanning electron microscopy.

Typically the specimens are prepared by cryofixation and freeze-fracture, after which the broken surface is freeze-etched, being eroded by exposure to vacuum till it shows the required level of detail. This technique can display protein molecules, organelle structures and lipid bilayers with very low degrees of distortion.

Condensation

Water vapour will only condense onto another surface when that surface is cooler than the dew point temperature, or when the water vapour equilibrium in air has been exceeded. When water vapour condenses onto a surface, a net warming occurs on that surface. The water molecule brings heat energy with it. In turn, the temperature of the atmosphere drops slightly. In the atmosphere, condensation produces clouds, fog and precipitation (usually only when facilitated by cloud condensation nuclei). The dew point of an air parcel is the temperature to which it must cool before water vapour in the air begins to condense.

Also, a net condensation of water vapour occurs on surfaces when the temperature of the surface is at or below the dew point temperature of the atmosphere. Deposition, the direct formation of ice from water vapour, is a type of condensation. Frost and snow are examples of deposition.

Chemical Reactions

A number of chemical reactions have water as a product. If the reactions take place at temperatures higher than the dew point of the surrounding air the water will be formed as vapour and increase the local humidity, if below the dew point local condensation will occur. Typical reactions that result in water formation are the burning of hydrogen or many other hydrocarbons in air itself or in combination with oxygen or other oxidisers.

In a similar fashion other chemical or physical reactions can take place in the presence of water vapour resulting in new chemicals forming such as rust on iron or steel, polymerisation occurring (certain polyurethane foams and cyanoacrylate glues cure with exposure to atmospheric humidity) or forms changing such as where anhydrous chemicals may absorb enough vapour to form a crystalline structure or alter an existing one, sometimes resulting in characteristic colour changes that can be used for measurement.

Measurement

Measuring the quantity of water vapour in a medium can be done directly or remotely with varying degrees of accuracy. Remote methods such electromagnetic absorption are possible from satellites above planetary atmospheres. Direct methods may use electronic transducers, moistened thermometres or hygroscopic materials measuring changes in physical properties or dimensions.

	medium	*temperature range (degC)*	*measurement uncertainty*	*typical measurement frequency*	*system cost*	*notes*
sling psychrometer	air	– to 50	low to moderate	hourly	low	
satellite-based spectroscopy	air	–80 to 60	low		very high	
capacitive sensor	air/gases	–40 to 50	moderate	2 to 0.05 Hz	medium	prone to becoming saturated/contaminated over time
warmed capacitive sensor	air/gases	–15 to 50	moderate to low	2 to 0.05 Hz (temp dependant)	medium to high	prone to becoming saturated/contaminated over time
resistive sensor	air/gases	–10 to 50	moderate	60 seconds	medium	prone to contamination
lithium chloride dewcell	air	–30 to 50	moderate	continuous	medium	see dewcell
Cobalt(II) chloride	air/gases	0 to 50	high	5 minutes	very low	often used in Humidity indicator card
Absorption spectroscopy	air/gases		moderate		high	
Aluminum oxide	air/gases		moderate		medium	see Moisture analysis
silicon oxide	air/gases		moderate		medium	see Moisture analysis
Piezoelectric sorption	air/gases		moderate		medium	see Moisture analysis

Electrolytic	air/gases		moderate		medium	see Moisture analysis
hair tension	air	0 to 40	high	continuous	low to medium	Affected by temperature. Adversely affected by prolonged high concentrations
Nephelometer	air/other gases		low		very high	
Goldbeater's skin (cow Peritoneum)	air	–20 to 30	moderate (with corrections)	slow, slower at lower temperatures	low	ref:WMO Guide to Meteorological Instruments and Methods of Observation No. 8 2006, (pages 1.12–1)
Lyman-alpha				high frequency	high	http://amsglossary.allenpress.com/glossary/search?id=lyman-alpha-hygrometer1 Requires frequent calibration
Gravimetric Hygrometer			very low		very high	often called primary source, national independent standards developed in US,UK,EU & Japan
	medium	temperature range (degC)	measurement uncertainty	typical measurement frequency	system cost	notes

Water Vapour Density

Water vapour is lighter or less dense than dry air. At equivalent temperatures it is buoyant with respect to dry air, whereby the density of dry air at standard temperature and pressure is 1.27 g/l and water vapour at standard temperature and pressure has the much lower density of .804 g/l.

Water Vapour and Dry Air Density Calculations at 0 °C

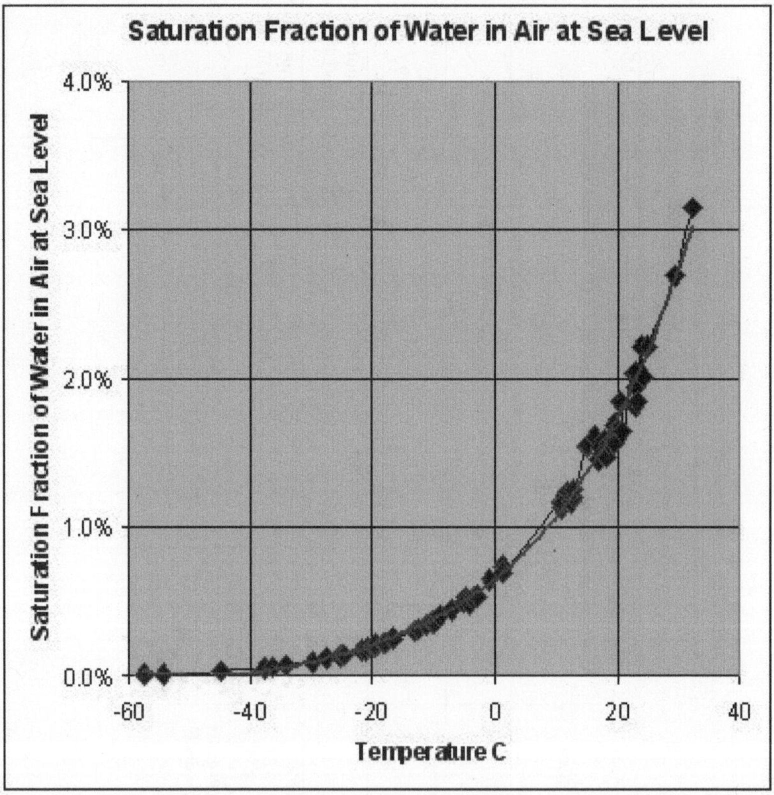

Figure: *The molar mass of water is 18.02 g/mol, as calculated from the sum of the atomic masses of its constituent atoms.*

The average molecular mass of air (approx. 79% nitrogen, N_2; 21% oxygen, O_2) is 28.57 g/mol at standard temperature and pressure (STP).

Using Avogadro's Law and the ideal gas law, water vapour and air will have a molar volume of 22.414 L/mol at STP. A molar mass of air and water vapour occupy the same volume of 22.414 litres. The density (mass/volume) of water vapour is 0.804 g/L, which is significantly less than that of dry air at 1.27 g/L at STP. This means water vapour is lighter than air.

STP conditions imply a temperature of 0 °C, at which the ability of water to become vapour is very restricted. Its concentration in air is very low at 0 °C. The red line on the chart to the right is the maximum concentration of water vapour expected for a given temperature. The water vapour concentration increases significantly as the temperature rises, approaching 100% (steam, pure water vapour) at 100 °C. However the difference in densities between air and water vapour would still exist.

Air and Water Vapour Density Interactions at Equal Temperatures

At the same temperature, a column of dry air will be denser or heavier than a column of air containing any water vapour. Thus, any volume of dry air will sink if placed in a larger volume of moist air. Also, a volume of moist air will rise or be buoyant if placed in a larger region of dry air. As the temperature rises the proportion of water vapour in the air increases, and its buoyancy will increase. The increase in buoyancy can have a significant atmospheric impact, giving rise to powerful, moisture rich, upward air currents when the air temperature and sea temperature reaches 25 °C or above. This phenomenon provides a significant motivating force for cyclonic and anticyclonic weather systems (typhoons and hurricanes).

Water Vapour and Respiration or Breathing

Water vapour is a by-product of respiration in plants and animals. Its contribution to the pressure, increases as its concentration increases. Its partial pressure contribution to air pressure increases, lowering the partial pressure contribution of the other atmospheric gases (Dalton's Law). The total air pressure must remain constant. The presence of water vapour in the air naturally dilutes or displaces the other air components as its concentration increases.

This can have an effect on respiration. In very warm air (35 °C) the proportion of water vapour is large enough to give rise to the stuffiness that can be experienced in humid jungle conditions or in poorly ventilated buildings.

Lifting Gas

Water vapour has lower density than that of air and is therefore buoyant in air but has lower vapour pressure than that of air. When water vapour is used as a lifting gas for use by a thermal airship the water vapour is heated to form steam so that its vapour pressure is greater than the surrounding air pressure in order to pressurize and to maintain the shape a theoretical "steam balloon", which yields approximately 60% the lift of helium and twice that of hot air.

General Discussion

The amount of water vapour in an atmosphere is constrained by the restrictions of partial pressures and temperature. Dew point temperature and relative humidity act as guidelines for the process of water vapour in the water cycle. Energy input, such as sunlight, can trigger more evaporation on an ocean surface or more sublimation on a chunk of ice on top of a mountain. The *balance* between condensation and evaporation gives the quantity called vapour partial pressure.

The maximum partial pressure (*saturation pressure*) of water vapour in air varies with temperature of the air and water vapour mixture. A variety of empirical formulas exist for this quantity; the most used reference formula is the Goff-Gratch equation for the SVP over liquid water below zero degree Celsius:

$\log_{10}(p) =$	$-7.90298\left(\frac{373.16}{T}-1\right)+5.02808\log_{10}\frac{373.16}{T}$
	$-1.3816\times10^{-7}\left(10^{11.344\left(1-\frac{T}{373.16}\right)}-1\right)$
	$+8.1328\times10^{-3}\left(10^{-3.49149\left(\frac{373.16}{T}-1\right)}-1\right)$
	$+\log_{10}(1013.246)$

Where T, temperature of the moist air, is given in units of kelvins, and p is given in units of millibars (hectopascals).

The formula is valid from about –50 to 102 °C; however there are a very limited number of measurements of the vapour pressure of water over supercooled liquid water. There are a number of other formulae which can be used.

Under certain conditions, such as when the boiling temperature of water is reached, a net evaporation will always occur during standard atmospheric conditions regardless of the percent of relative humidity. This immediate process will dispel massive amounts of water vapour into a cooler atmosphere.

Exhaled air is almost fully at equilibrium with water vapour at the body temperature. In the cold air the exhaled vapour quickly condenses, thus showing up as a fog or mist of water droplets and as condensation or frost on surfaces. Forcibly condensing these water droplets from exhaled breath is the basis of exhaled breath condensate, an evolving medical diagnostic test. Controlling water vapour in air is a key concern in the heating, ventilating, and air-conditioning (HVAC) industry. Thermal comfort depends on the moist air conditions. Non-human comfort situations are called refrigeration, and also are affected by water vapour. For example many food stores, like supermarkets, utilize open chiller cabinets, or *food cases*, which can significantly lower the water vapour pressure (lowering humidity). This practice delivers several benefits as well as problems.

Water Vapour in Earth's Atmosphere

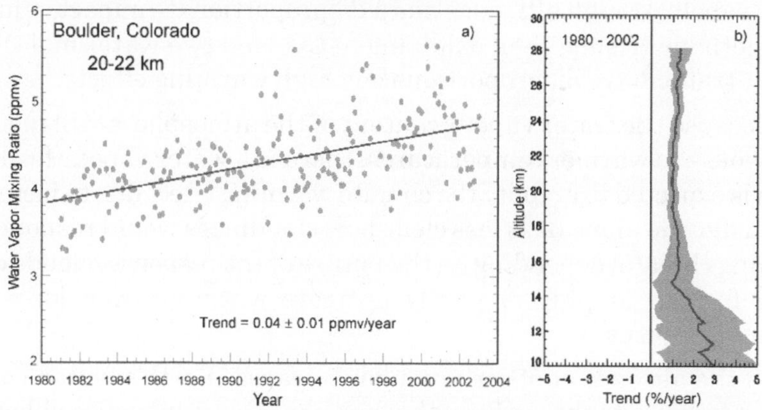

Figure: *Evidence for increasing amounts of stratospheric water vapour over time in Boulder, Colorado.*

Gaseous water represents a small but environmentally significant constituent of the atmosphere. The percentage water vapour in surface air varies from .01% at -42 °C (-44 °F) to 4.24% when the dew point is 30 °C (86 °F). Approximately 99.13% of it is contained in the troposphere. The condensation of water vapour to the liquid or ice phase is responsible for clouds, rain, snow, and other precipitation, all of which count among the most significant elements of what we experience as weather. Less obviously, the latent heat of vaporization, which is released to the atmosphere whenever condensation occurs, is one of the most important terms in the atmospheric energy budget on both local and global scales. For example, latent heat release in atmospheric convection is directly responsible for powering destructive storms such as tropical cyclones and severe thunderstorms. Water vapour is also the most potent greenhouse gas owing to the presence of the hydroxyl bond which strongly absorbs in the infra-red region of the light spectrum.

Water in Earth's atmosphere is not merely below its boiling point (100°C), but at altitude it goes below its freezing point (0°C), due to water's highly polar attraction. When combined with its quantity, water vapour then has a relevant dew point and frost point, unlike e. g., carbon dioxide and methane.

Water vapour thus has a scale height a fraction of that of the bulk atmosphere, as the water condenses and exits, primarily in the troposphere, the lowest layer of the atmosphere. Carbon dioxide and methane, being non-polar, rise above water vapour. The absorption and emission of CO_2 and CH_4 are then significant to Earth's emission to space, and thus the planetary greenhouse effect. Conversely, adding water vapour at high altitudes has a disproportionate impact, which is why methane (rising, then oxidizing to CO_2 and two water molecules) and jet traffic have disproportionately high warming effects.

Because the water vapour content of the atmosphere will increase in response to warmer temperatures, there is a water vapour feedback which is expected to amplify the climate warming effect due to increased carbon dioxide alone. It is less clear how cloudiness would respond to a warming climate; depending on the nature of the response, clouds could either further amplify or partly mitigate warming from long-lived greenhouse gases.

In the absence of other greenhouse gases, Earth's water vapour would condense to the surface; this has likely happened, possibly more than once. Scientists thus distinguish between non-condensable (driving) and condensable (driven) greenhouse gases- i.e., the above water vapour feedback.

Fog and clouds form through condensation around cloud condensation nuclei. In the absence of nuclei, condensation will only occur at much lower temperatures. Under persistent condensation or deposition, cloud droplets or snowflakes form, which precipitate when they reach a critical mass.

The water content of the atmosphere as a whole is constantly depleted by precipitation. At the same time it is constantly replenished by evaporation, most prominently from seas, lakes, rivers, and moist earth. Other sources of atmospheric water include combustion, respiration, volcanic eruptions, the transpiration of plants, and various other biological and geological processes. The mean global content of water vapour in the atmosphere is roughly sufficient to cover the surface of the planet with a layer of liquid water about 25 mm deep. The mean annual precipitation for the planet is about 1 metre, which implies a rapid turnover of water in the air – on average, the residence time of a water molecule in the troposphere is about 9 to 10 days.

Episodes of surface geothermal activity, such as volcanic eruptions and geysers, release variable amounts of water vapour into the atmosphere. Such eruptions may be large in human terms, and major explosive eruptions may inject exceptionally large masses of water exceptionally high into the atmosphere, but as a percentage of total atmospheric water, the role of such processes is minor. The relative concentrations of the various gases emitted by volcanoes varies considerably according to the site and according to the particular event at any one site. However, water vapour is consistently the commonest volcanic gas; as a rule, it comprises more than 60% of total emissions during a subaerial eruption.

Atmospheric water vapour content is expressed using various measures. These include vapour pressure, specific humidity, mixing ratio, dew point temperature, and relative humidity.

Radar and Satellite Imaging

Because water molecules absorb microwaves and other radio wave frequencies, water in the atmosphere attenuates radar signals. In addition, atmospheric water will reflect and refract signals to an extent that depends on whether it is vapour, liquid or solid.

Generally, radar signals lose strength progressively the farther they travel through the troposphere. Different frequencies attenuate at different rates, such that some components of air are opaque to some frequencies and transparent to others. Radio waves used for broadcasting and other communication experience the same effect.

Water vapour reflects radar to a lesser extent than do water's other two phases. In the form of drops and ice crystals, water acts as a prism, which it does not do as an individual molecule; however, the existence of water vapour in the atmosphere causes the atmosphere to act as a giant prism.

A comparison of GOES-12 satellite images shows the distribution of atmospheric water vapour relative to the oceans, clouds and continents of the Earth. Vapour surrounds the planet but is unevenly distributed.

Lightning Generation

Water vapour plays a key role in lightning production in the atmosphere. From cloud physics, usually, clouds are the real generators of static charge as found in Earth's atmosphere. But the ability, or capability of clouds to hold massive amounts of electrical energy is directly related to the amount of water vapour present in the local system.

The amount of water vapour directly controls the permittivity of the air. During times of low humidity, static discharge is quick and easy. During times of higher humidity, fewer static discharges occur. Permittivity and capacitance work hand in hand to produce the megawatt outputs of lightning.

After a cloud, for instance, has started its way to becoming a lightning generator, atmospheric water vapour acts as a substance (or insulator) that decreases the ability of the cloud to discharge its electrical energy. Over a certain amount of time, if the cloud continues to generate and store more static electricity, the barrier that was created by the atmospheric water vapour will ultimately break down from the stored electrical potential energy. This energy will be released to a locally, oppositely charged region in the form of lightning. The strength of each discharge is directly related to the atmospheric permittivity, capacitance, and the source's charge generating ability.

Chapter 5

Ozone

Ozone (systematically named $1\lambda^1,3\lambda^1$-trioxidane and *catena*-trioxygen), or trioxygen, is an inorganic molecule with the chemical formula OO_2 (also written $[O_3]$). It is a pale blue gas with a distinctively pungent smell. It is an allotrope of oxygen that is much less stable than the diatomic allotrope O_2, breaking down in the lower atmosphere to normal dioxygen. Ozone is formed from dioxygen by the action of ultraviolet light and also atmospheric electrical discharges, and is present in low concentrations throughout the Earth's atmosphere. In total, ozone makes up only 0.6 ppm of the atmosphere.

Ozone's odor is sharp, reminiscent of chlorine, and detectable by many people at concentrations of as little as 10 ppb in air. Ozone's O_3 formula was determined in 1865. The molecule was later proven to have a bent structure and to be diamagnetic. In standard conditions, ozone is a pale blue gas that condenses at progressively cryogenic temperatures to a dark blue liquid and finally a violet-black solid. Ozone's instability with regard to more common dioxygen is such that both concentrated gas and liquid ozone may decompose explosively. It is therefore used commercially only in low concentrations.

Ozone is a powerful oxidant (far more so than dioxygen) and has many industrial and consumer applications related to oxidation. This same high oxidizing potential, however, causes ozone to damage mucous and respiratory tissues in animals, and also tissues in plants, above concentrations of about 100 ppb. This makes ozone a potent respiratory hazard and pollutant near ground level.

However, the so-called ozone layer (a portion of the stratosphere with a higher concentration of ozone, from two to eight ppm) is beneficial, preventing damaging ultraviolet light from reaching the Earth's surface, to the benefit of both plants and animals.

Nomenclature

The trivial name *ozone* is the most commonly used and preferred IUPAC name. The systematic names *1λ¹,3λ¹-trioxidane* and *catena-trioxygen*, valid IUPAC names, are constructed according to the substitutive and additive nomenclatures, respectively. The name *ozone* derives from *ozein*, the Greek word for smell (verb), referring to ozone's distinctive smell.

In appropriate contexts, ozone can be viewed as trioxidane with two hydrogen atoms removed, and as such, *trioxidanylidene* may be used as a context-specific systematic name, according to substitutive nomenclature. By default, these names pay no regard to the radicality of the ozone molecule. In even more specific context, this can also name the non-radical singlet ground state, whereas the diradical state is named *trioxidanediyl*.

Trioxidanediyl (or *ozonide*) is used, non-systematically, to refer to the substituent group (-OOO-). Care should be taken to avoid confusing the name of the group for the context-specific name for ozone given above.

History

Ozone, the first allotrope of any chemical element to be recognized, was proposed as a distinct chemical substance by Christian Friedrich Schönbein in 1840, who named it after the Greek verb ozein (Dæåév, "to smell"), from the peculiar odor in lightning storms. The formula for ozone, O_3, was not determined until 1865 by Jacques-Louis Soret and confirmed by Schönbein in 1867.

For much of the second half of the nineteenth century and well into the twentieth, ozone was considered a healthy component of the environment by naturalists and health-seekers. The City of Beaumont in California had as its official slogan "Beaumont: Zone of Ozone," as evidenced on postcards and Chamber of Commerce letterhead. Naturalists working outdoors often considered the higher elevations beneficial because of their ozone content. "There is quite a different atmosphere [at higher elevation] with enough ozone to sustain the necessary energy [to work]," wrote naturalist Henry Henshaw, working in Hawaii. Seaside air was considered to be healthy because of its "ozone" content but the smell giving rise to this belief is in reality that of rotting seaweed.

Physical Properties

Ozone is colourless or slightly bluish gas (blue when liquified), slightly soluble in water and much more soluble in inert non-polar solvents such as carbon tetrachloride or fluorocarbons, where it forms a blue solution. At 161 K (–112 °C; –170 °F), it condenses to form a dark blue liquid.

It is dangerous to allow this liquid to warm to its boiling point, because both concentrated gaseous ozone and liquid ozone can detonate. At temperatures below 80 K (–193.2 °C; –315.7 °F), it forms a violet-black solid.

Most people can detect about 0.01 μmol/mol of ozone in air where it has a very specific sharp odor somewhat resembling chlorine bleach. Exposure of 0.1 to 1 μmol/mol produces headaches, burning eyes and irritation to the respiratory passages.

Even low concentrations of ozone in air are very destructive to organic materials such as latex, plastics and animal lung tissue.

Ozone is diamagnetic, which means that its electrons are all paired. In contrast, O_2 is paramagnetic, containing two unpaired electrons.

Structure

According to experimental evidence from microwave spectroscopy, ozone is a bent molecule, with C_{2v} symmetry (similar to the water molecule). The O – O distances are 127.2 pm (1.272 Å).

The O – O – O angle is 116.78°. The central atom is sp^2 hybridized with one lone pair. Ozone is a polar molecule with a dipole moment of 0.53 D. The bonding can be expressed as a resonance hybrid with a single bond on one side and double bond on the other producing an overall bond order of 1.5 for each side.

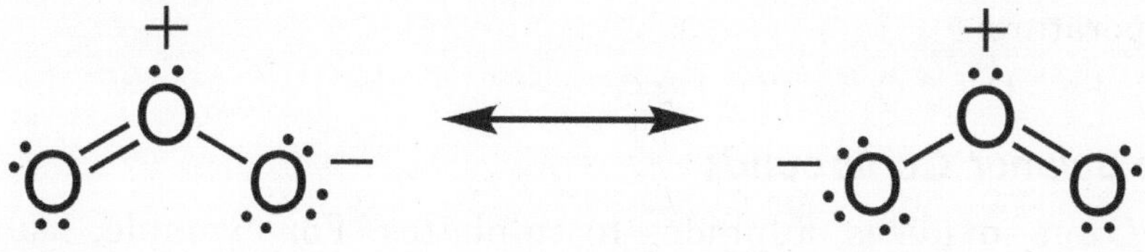

Reactions

Ozone is a powerful oxidizing agent, far stronger than O_2. It is also unstable at high concentrations, decaying to ordinary diatomic oxygen. It has a varying half-life length, depending upon atmospheric conditions (temperature, humidity, and air movement). In a sealed chamber with a fan that moves the gas, ozone has a half-life of approximately a day at room temperature.

Some unverified claims imply that ozone can have a half life as short as a half an hour under atmospheric conditions.

$$2\,O_3 \rightarrow 3\,O_2$$

This reaction proceeds more rapidly with increasing temperature and increased pressure. Deflagration of ozone can be triggered by a spark, and can occur in ozone concentrations of 10 wt% or higher.

With Metals

Ozone will oxidize most metals (except gold, platinum, and iridium) to oxides of the metals in their highest oxidation state. For example:

$$2\,Cu^{+} + 2\,H_3O^{+} + O_3 \rightarrow 2\,Cu^{2+} + 3\,H_2O + O_2$$

With Nitrogen and Carbon Compounds

Ozone also oxidizes nitric oxide to nitrogen dioxide:

$$NO + O_3 \rightarrow NO_2 + O_2$$

This reaction is accompanied by chemiluminescence. The NO_2 can be further oxidized:

$$NO_2 + O_3 \rightarrow NO_3 + O_2$$

The NO_3 formed can react with NO_2 to form N_2O_5.

Solid nitronium perchlorate can be made from NO_2, ClO_2, and O_3 gases:

$$2\,NO_2 + 2\,ClO_2 + 2\,O_3 \rightarrow 2\,NO_2ClO_4 + O_2$$

Ozone does not react with ammonium salts, but it oxidizes ammonia to ammonium nitrate:

$$2\,NH_3 + 4\,O_3 \rightarrow NH_4NO_3 + 4\,O_2 + H_2O$$

Ozone reacts with carbon to form carbon dioxide, even at room temperature:

$$C + 2\,O_3 \rightarrow CO_2 + 2\,O_2$$

With Sulphur Compounds

Ozone oxidizes sulphides to sulphates. For example, lead(II) sulphide is oxidised to lead(II) sulphate:

$$PbS + 4\,O_3 \rightarrow PbSO_4 + 4\,O_2$$

Sulphuric acid can be produced from ozone, water and either elemental sulphur or sulphur dioxide:

$$S + H_2O + O_3 \rightarrow H_2SO_4$$

$$3\,SO_2 + 3\,H_2O + O_3 \rightarrow 3\,H_2SO_4$$

In the gas phase, ozone reacts with hydrogen sulfide to form sulphur dioxide:

$$H_2S + O_3 \rightarrow SO_2 + H_2O$$

In an aqueous solution, however, two competing simultaneous reactions occur, one to produce elemental sulphur, and one to produce sulphuric acid:

$$H_2S + O_3 \rightarrow S + O_2 + H_2O$$

$$3\ H_2S + 4\ O_3 \rightarrow 3\ H_2SO_4$$

With Alkenes and Alkynes

Alkenes can be oxidatively cloven by ozone, in a process called ozonolysis, giving alcohols, aldehydes, ketones, and carboxylic acids, depending on the second step of the workup.

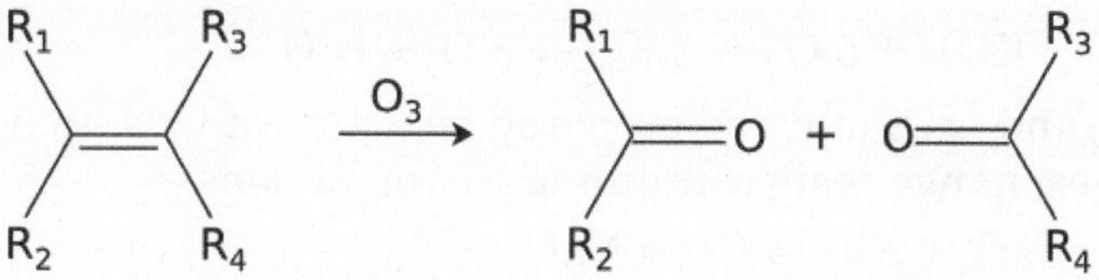

Usually ozonolysis is carried out in a solution of dichloromethane, at a temperature of –78°C. After a sequence of cleavage and rearrangement, an organic ozonide is formed. With reductive workup (e.g. zinc in acetic acid or dimethyl sulfide), ketones and aldehydes will be formed, with oxidative workup (e.g. aqueous or alcoholic hydrogen peroxide), carboxylic acids will be formed.

Other Substrates

All three atoms of ozone may also react, as in the reaction of tin(II) chloride with hydrochloric acid and ozone:

$$3\ SnCl_2 + 6\ HCl + O$$

$$3 \rightarrow 3\ SnCl_4 + 3\ H_2O$$

Iodine perchlorate can be made by treating iodine dissolved in cold anhydrous perchloric acid with ozone:

$$I_2 + 6\ HClO_4 + O_3 \rightarrow 2\ I(ClO_4)_3 + 3\ H_2O$$

Combustion

Ozone can be used for combustion reactions and combustible gases; ozone provides higher temperatures than burning in dioxygen (O_2). The following is a reaction for the combustion of carbon subnitride which can also cause higher temperatures:

$$3\,C_4N_2 + 4\,O_3 \rightarrow 12\,CO + 3\,N_2$$

Ozone can react at cryogenic temperatures. At 77 K (–196.2 °C; –321.1 °F), atomic hydrogen reacts with liquid ozone to form a hydrogen superoxide radical, which dimerizes:

$$H + O_3 \rightarrow HO_2 + O2\,HO_2 \rightarrow H_2O_4$$

Reduction to Ozonides

Reduction of ozone gives the ozonide anion, O^-3. Derivatives of this anion are explosive and must be stored at cryogenic temperatures. Ozonides for all the alkali metals are known. KO_3, RbO_3, and CsO_3 can be prepared from their respective superoxides:

$$KO_2 + O_3 \rightarrow KO_3 + O_2$$

Although KO_3 can be formed as above, it can also be formed from potassium hydroxide and ozone:

$$2\,KOH + 5\,O_3 \rightarrow 2\,KO_3 + 5\,O_2 + H_2O$$

NaO_3 and LiO_3 must be prepared by action of CsO_3 in liquid NH_3 on an ion exchange resin containing Na^+ or Li^+ ions:

$$CsO_3 + Na^+ \rightarrow Cs^+ + NaO_3$$

A solution of calcium in ammonia reacts with ozone to give to ammonium ozonide and not calcium ozonide:

$$3\,Ca + 10\,NH_3 + 6\,O$$

$$3 \rightarrow Ca{\cdot}6NH_3 + Ca(OH)_2 + Ca(NO_3)_2 + 2\,NH_4O_3 + 2\,O_2 + H_2$$

Applications

Ozone can be used to remove manganese from water, forming a precipitate which can be filtered:

$$2\,Mn^{2+} + 2\,O_3 + 4\,H_2O \rightarrow 2\,MnO(OH)_2\,(s) + 2\,O_2 + 4\,H^+$$

Ozone will also detoxify cyanides by converting them to cyanates, which are a thousand times less toxic.

$$CN^- + O_3 \rightarrow CNO^- + O_2$$

Ozone will also completely decompose urea:

$$(NH_2)_2CO + O_3 \rightarrow N_2 + CO_2 + 2\,H_2O$$

Ozone in Earth's Atmosphere

The standard way to express total ozone levels (the amount of ozone in a vertical column) in the atmosphere is by using Dobson units. Point measurements are reported as mole fractions in nmol/mol (parts per billion, ppb) or as concentrations in $\mu g/m^3$. The study of ozone concentration in the atmosphere started in the 1920s.

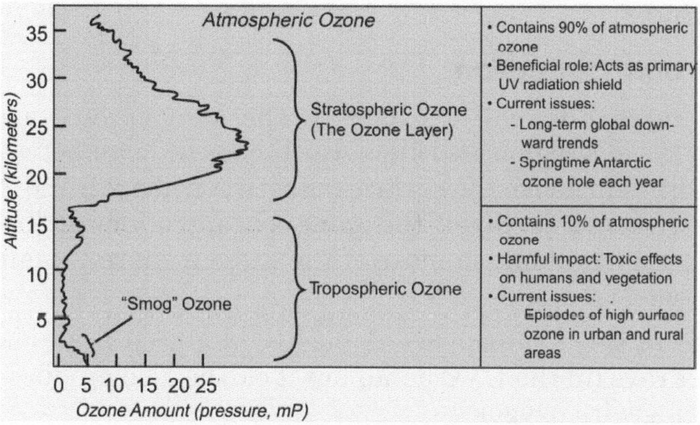

Figure: *The distribution of atmospheric ozone in partial pressure as a function of altitude*

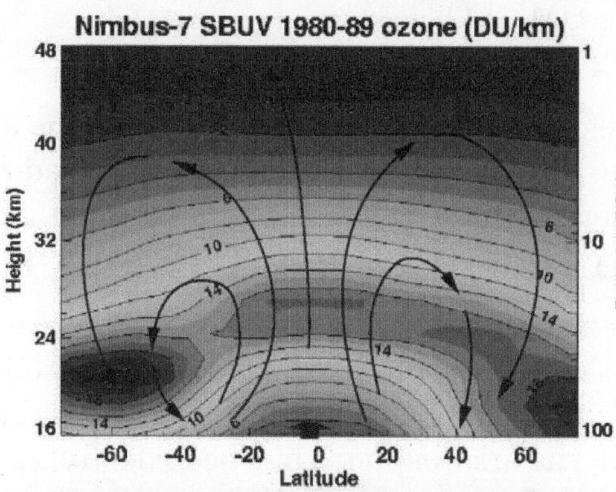

Figure: *Concentration of ozone as measured by the Nimbus-7 satellite*

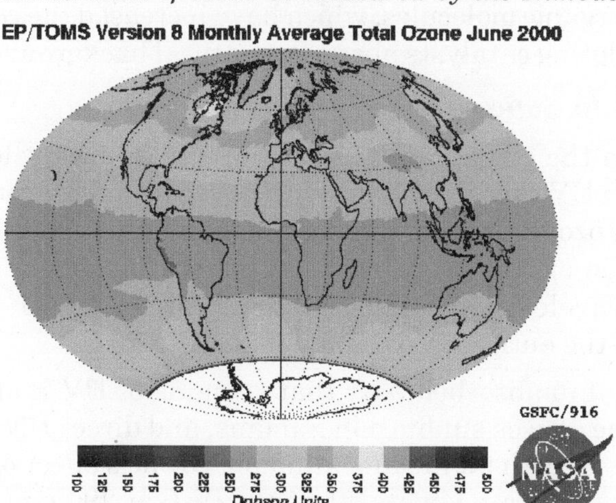

Figure: *Total ozone concentration in June 2000 as measured by EP-TOMS satellite instrument*

Ozone Layer

Location and Production

The highest levels of ozone in the atmosphere are in the stratosphere, in a region also known as the ozone layer between about 10 km and 50 km above the surface (or between about 6 and 31 miles). However, even in this "layer" the ozone concentrations are only two to eight parts per million, so most of the oxygen there remains of the dioxygen type.

Ozone in the stratosphere is mostly produced from short-wave ultraviolet rays (in the UVC band) but it can be also produced from x-rays reacting with oxygen:

$$O_2 + \text{photon (radiation } \lambda < 240 \text{ nm)} \rightarrow 2\,O$$

$$O + O_2 + M \rightarrow O_3 + M$$

$$\alpha + \beta^- + O_2 \rightarrow He + O_3$$

where "M" denotes the third body that carries off the excess energy of the reaction. The thus produced ozone is destroyed by the reaction with atomic oxygen:

$$O3 + O \rightarrow 2\,O_2$$

The latter reaction is catalysed by the presence of certain free radicals, of which the most important are hydroxyl (OH), nitric oxide (NO) and atomic chlorine (Cl) and bromine (Br). In recent decades the amount of ozone in the stratosphere has been declining mostly because of emissions of chlorofluorocarbons (CFC) and similar chlorinated and brominated organic molecules, which have increased the concentration of ozone-depleting catalysts above the natural background.

Importance to Surface-dwelling Life on Earth

Ozone in the ozone layer filters out sunlight wavelengths from about 200 nm UV rays to 315 nm, with ozone peak absorption at about 250 nm. This ozone UV absorption is important to life, since it extends the absorption of UV by ordinary oxygen and nitrogen in air (which absorb all wavelengths < 200 nm) through the lower UV-C (200–280 nm) and the entire UV-B band (280–315 nm).

The small unabsorbed part that remains of UV-B after passage through ozone causes sunburn in humans, and direct DNA damage in living tissues in both plants and animals. Ozone's effect on mid-range UV-B rays is illustrated by its effect on UV-B at 290 nm, which has a radiation intensity 350 million times as powerful at the top of the atmosphere as at the surface.

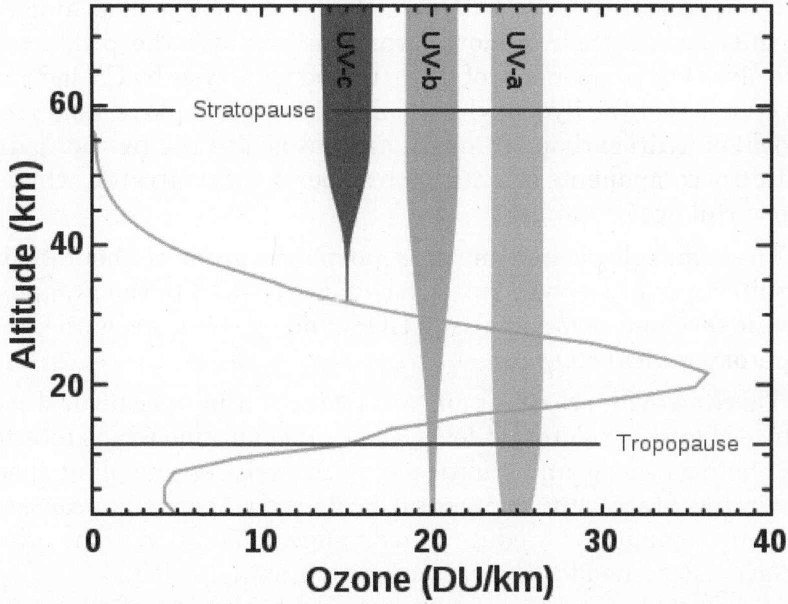

Figure: *Levels of ozone at various altitudes and blocking of different bands of ultraviolet radiation. Essentially all UVC (100–280 nm) is blocked by dioxygen (at 100–200 nm) or by ozone (at 200–280 nm) in the atmosphere. The shorter portion of this band and even more energetic UV causes the formation of the ozone layer, when single oxygen atoms produced by UV photolysis of dioxygen (below 240 nm) react with more dioxygen. The ozone layer itself then blocks most, but not quite all, sunburn-producing UVB (280–315 nm). The band of UV closest to visible light, UVA (315–400 nm), is hardly affected by ozone, and most of it reaches the ground.*

Nevertheless, enough of UV-B radiation at similar frequency reaches the ground to cause some sunburn, and these same wavelengths are also among those responsible for the production of vitamin D in humans.

The ozone layer has little effect on the longer UV wavelengths called UV-A (315–400 nm), but this radiation does not cause sunburn or direct DNA damage, and while it probably does cause long-term skin damage in certain humans, it is not as dangerous to plants and to the health of surface-dwelling organisms on Earth in general.

Low Level Ozone

Low level ozone (or tropospheric ozone) is an atmospheric pollutant. It is not emitted directly by car engines or by industrial operations, but formed by the reaction of sunlight on air containing hydrocarbons and nitrogen oxides that react to form ozone directly at the source of the pollution or many kilometres down wind.

Ozone reacts directly with some hydrocarbons such as aldehydes and thus begins their removal from the air, but the products are themselves key components of smog. Ozone photolysis by UV light leads to production of the hydroxyl radical HO• and this plays a part in the removal of hydrocarbons from the air, but is also the first step in the creation of components of smog such as peroxyacyl nitrates, which can be powerful eye irritants.

The atmospheric lifetime of tropospheric ozone is about 22 days; its main removal mechanisms are being deposited to the ground, the above-mentioned reaction giving HO•, and by reactions with OH and the peroxy radical HO_2•.

There is evidence of significant reduction in agricultural yields because of increased ground-level ozone and pollution which interferes with photosynthesis and stunts overall growth of some plant species. The United States Environmental Protection Agency is proposing a secondary regulation to reduce crop damage, in addition to the primary regulation designed for the protection of human health.

Certain examples of cities with elevated ozone readings are Houston, Texas, and Mexico City, Mexico. Houston has a reading of around 41 nmol/mol, while Mexico City is far more hazardous, with a reading of about 125 nmol/mol.

Ozone Cracking

Ozone gas attacks any polymer possessing olefinic or double bonds within its chain structure, such as natural rubber, nitrile rubber, and styrene-butadiene rubber. Products made using these polymers are especially susceptible to attack, which causes cracks to grow longer and deeper with time, the rate of crack growth depending on the load carried by the rubber component and the concentration of ozone in the atmosphere. Such materials can be protected by adding antiozonants, such as waxes, which bond to the surface to create a protective film or blend with the material and provide long term protection.

Ozone cracking used to be a serious problem in car tires for example, but the problem is now seen only in very old tires. On the other hand, many critical products like gaskets and O-rings may be attacked by ozone produced within compressed air systems. Fuel lines made of reinforced rubber are also susceptible to attack, especially within the engine compartment, where some ozone is produced by electrical components. Storing rubber products in close proximity to a DC electric motor can accelerate ozone cracking. The commutator of the motor generates sparks which in turn produce ozone.

Ozone as a Greenhouse Gas

Although ozone was present at ground level before the Industrial Revolution, peak concentrations are now far higher than the pre-industrial levels, and even background concentrations well away from sources of pollution are substantially higher. Ozone acts as a greenhouse gas, absorbing some of the infrared energy emitted by the earth. Quantifying the greenhouse gas potency of ozone is difficult because it is not present in uniform concentrations across the globe. However, the most widely accepted scientific assessments relating to climate change (e.g. the Intergovernmental Panel on Climate Change Third Assessment Report) suggest that the radiative forcing of tropospheric ozone is about 25% that of carbon dioxide.

The annual global warming potential of tropospheric ozone is between 918–1022 tons carbon dioxide equivalent/tons tropospheric ozone. This means on a per-molecule basis, ozone in the troposphere has a radiative forcing effect roughly 1,000 times as strong as carbon dioxide. However, tropospheric ozone is a short-lived greenhouse gas, which decays in the atmosphere much more quickly than carbon dioxide. This means that over a 20 year span, the global warming potential of tropospheric ozone is much less, roughly 62 to 69 tons carbon dioxide equivalent / ton tropospheric ozone.

Because of its short-lived nature, tropospheric ozone does not have strong global effects, but has very strong radiative forcing effects on regional scales. In fact, there are regions of the world where tropospheric ozone has a radiative forcing up to 150% of carbon dioxide.

Production

Ozone often forms in nature under conditions where O_2 will not react. Ozone used in industry is measured in μmol/mol (ppm, parts per million), nmol/mol (ppb, parts per billion), $\mu g/m^3$, mg/h (milligrams per hour) or weight percent. The regime of applied concentrations ranges from 1 to 5% in air and from 6 to 14% in oxygen for older generation methods. New electrolytic methods can achieve up 20 to 30% dissolved ozone concentrations in output water.

Temperature and humidity plays a large role in how much ozone is being produced using traditional generation methods such as corona discharge and ultraviolet light. Old generation methods will produce less than 50% its nominal capacity if operated with humid ambient air than when it operates in very dry air. New generators using electrolytic methods can achieve higher purity and dissolution through using water molecules as the source of ozone production.

Corona Discharge Method

This is the most common type of ozone generator for most industrial and personal uses. While variations of the "hot spark" coronal discharge method of ozone production exist, including medical grade and industrial grade ozone generators, these units usually work by means of a corona discharge tube. They are typically cost-effective and do not require an oxygen source other than the ambient air to produce ozone concentrations of 3–6%. Fluctuations in ambient air, due to weather or other environmental conditions, cause variability in ozone production. However, they also produce nitrogen oxides as a by-product. Use of an air dryer can reduce or eliminate nitric acid formation by removing water vapour and increase ozone production. Use of an oxygen concentrator can further increase the ozone production and further reduce the risk of nitric acid formation by removing not only the water vapour, but also the bulk of the nitrogen.

Ultraviolet Light

UV ozone generators, or vacuum-ultraviolet (VUV) ozone generators, employ a light source that generates a narrow-band ultraviolet light, a subset of that produced by the Sun. The Sun's UV sustains the ozone layer in the stratosphere of Earth.

While standard UV ozone generators tend to be less expensive, they usually produce ozone with a concentration of about 0.5% or lower. Another disadvantage of this method is that it requires the air (oxygen) to be exposed to the UV source for a longer amount of time, and any gas that is not exposed to the UV source will not be treated. This makes UV generators impractical for use in situations that deal with rapidly moving air or water streams (in-duct air sterilization, for example). Production of ozone is one of the potential dangers of ultraviolet germicidal irradiation. VUV ozone generators are used in swimming pool and spa applications ranging to millions of gallons of water. VUV ozone generators, unlike corona discharge generators, do not produce harmful nitrogen by-products and also unlike corona discharge systems, VUV ozone generators work extremely well in humid air environments. There is also not normally a need for expensive off-gas mechanisms, and no need for air driers or oxygen concentrators which require extra costs and maintenance.

Cold Plasma

In the cold plasma method, pure oxygen gas is exposed to a plasma created by dielectric barrier discharge. The diatomic oxygen is split into single atoms, which then recombine in triplets to form ozone.

Cold plasma machines utilize pure oxygen as the input source and produce a maximum concentration of about 5% ozone. They produce far greater quantities of ozone in a given space of time compared to ultraviolet production. However, because cold plasma ozone generators are very expensive, they are found less frequently than the previous two types.

The discharges manifest as filamentary transfer of electrons (micro discharges) in a gap between two electrodes. In order to evenly distribute the micro discharges, a dielectric insulator must be used to separate the metallic electrodes and to prevent arcing.

Some cold plasma units also have the capability of producing short-lived allotropes of oxygen which include O_4, O_5, O_6, O_7, etc. These species are even more reactive than ordinary O_3.

Electrolytic

Electrolytic ozone generation (EOG) splits water molecules into H_2, O_2, and O_3. In most EOG methods, the hydrogen gas will be removed to leave oxygen and ozone as the only reaction products. Therefore, EOG can achieve higher dissolution in water without other competing gases found in corona discharge method, such as nitrogen gases present in ambient air. This method of generation can achieve concentrations of 20–30% and is independent of air quality because water is used as the source material. Production of ozone electrolytically is typically unfavourable because of the high overpotential required to produce ozone as compared to oxygen. This is why ozone is not produced during typical water electrolysis. However, it is possible to increase the overpotential of oxygen by careful catalyst selection such that ozone is preferentially produced under electrolysis. Catalysts typically chosen for this approach are lead dioxide or boron-doped diamond.

Special Considerations

Ozone cannot be stored and transported like other industrial gases (because it quickly decays into diatomic oxygen) and must therefore be produced on site. Available ozone generators vary in the arrangement and design of the high-voltage electrodes. At production capacities higher than 20 kg per hour, a gas/water tube heat-exchanger may be utilized as ground electrode and assembled with tubular high-voltage electrodes on the gas-side. The regime of typical gas pressures is around 2 bars (200 kPa) absolute in oxygen and 3 bars (300 kPa) absolute in air. Several megawatts of electrical power may be installed in large facilities, applied as one phase AC current at 50 to 8000 Hz and peak voltages between 3,000 and 20,000 volts. Applied voltage is usually inversely related to the applied frequency.

The dominating parametre influencing ozone generation efficiency is the gas temperature, which is controlled by cooling water temperature and/or gas velocity. The cooler the water, the better the ozone synthesis. The lower the gas velocity, the higher the concentration (but the lower the net ozone produced). At typical industrial conditions, almost 90% of the effective power is dissipated as heat and needs to be removed by a sufficient cooling water flow.

Because of the high reactivity of ozone, only a few materials may be used like stainless steel (quality 316L), titanium, aluminium (as long as no moisture is present), glass, polytetrafluorethylene, or polyvinylidene fluoride. Viton may be used with the restriction of constant mechanical forces and absence of humidity (humidity limitations apply depending on the formulation). Hypalon may be used with the restriction that no water come in contact with it, except for normal atmospheric levels. Embrittlement or shrinkage is the common mode of failure of elastomers with exposure to ozone. Ozone cracking is the common mode of failure of elastomer seals like O-rings.

Silicone rubbers are usually adequate for use as gaskets in ozone concentrations below 1 wt%, such as in equipment for accelerated aging of rubber samples.

Incidental Production

Ozone may be formed from O_2 by electrical discharges and by action of high energy electromagnetic radiation. Unsuppressed arcing breaks down the chemical bonds of the atmospheric oxygen surrounding the contacts [$O_2 \rightarrow 2O$]. Free radicals of oxygen in and around the arc recombine to create ozone [O_3]. Certain electrical equipment generate significant levels of ozone. This is especially true of devices using high voltages, such as ionic air purifiers, laser printers, photocopiers, tasers and arc welders. Electric motors using brushes can generate ozone from repeated sparking inside the unit. Large motors that use brushes, such as those used by elevators or hydraulic pumps, will generate more ozone than smaller motors. Ozone is similarly formed in the Catatumbo lightning storms phenomenon on the Catatumbo River in Venezuela, which helps to replenish ozone in the upper troposphere. It is the world's largest single natural generator of ozone, lending calls for it to be designated a UNESCO World Heritage Site.

Laboratory Production

In the laboratory, ozone can be produced by electrolysis using a 9 volt battery, a pencil graphite rod cathode, a platinum wire anode and a 3 molar sulphuric acid electrolyte. The half cell reactions taking place are:

$$3\ H_2O \rightarrow O_3 + 6\ H^+ + 6\ e^- \ (\Delta E^o = -1.53\ V)$$

$$6\ H^+ + 6\ e^- \rightarrow 3\ H_2 \ (\Delta E^o = 0\ V)$$

$$2\ H_2O \rightarrow O_2 + 4\ H^+ + 4\ e^- \ (\Delta E^o = -1.23\ V)$$

In the net reaction, three equivalents of water are converted into one equivalent of ozone and three equivalents of hydrogen. Oxygen formation is a competing reaction.

It can also be "prepared" by high voltage arc. This can be done with an apparatus consisting of two concentric glass tubes sealed together at the top, with in and out spigots at the top and bottom of the outer tube. The inner core should have a length of metal foil inserted into it connected to one side of the power source. The other side of the power source should be connected to another piece of foil wrapped around the outer tube. Dry O_2 should be run through the tube in one spigot. As the O_2 is run through one spigot into the apparatus and high voltage is applied to the foil leads, electricity will discharge between the dry dioxygen in the middle and form O_3 and O_2 out the other spigot. The reaction can be summarized as follows:

$$3\ O_2 \text{ — } \textit{electricity} \rightarrow 2\ O_3$$

Agriculture

Ozone application on freshly cut pineapple and banana shows increase in flavonoids and total phenol contents when exposure is up to 20 minutes. Decrease in ascorbic acid (one form of vitamin C) content is observed but the positive effect on total phenol content and flavonoids can overcome the negative effect. Tomatoes upon treatment with ozone shows an increase in β-carotene, lutein and lycopene. However, ozone application on strawberries in pre-harvest period shows decrease in ascorbic acid content.

Ozone facilitates the extraction of some heavy metals from soil using EDTA. EDTA forms strong, water-soluble coordination compounds with some heavy metals (Pb, Zn) thereby making it possible to dissolve them out from contaminated soil. If contaminated soil is pre-treated with ozone, the extraction efficacy of Pb, Am and Pu increases by 11.0–28.9%, 43.5% and 50.7% respectively.

Sulphur Dioxide

Sulphur dioxide (also sulphur dioxide) is the chemical compound with the formula SO_2. At standard atmosphere, it is a toxic gas with a pungent, irritating, and rotten smell. The triple point is 197.69 K and 1.67kPa. It is released naturally by volcanic activity.

Sulphur dioxide was used by the Romans in winemaking, when they discovered that burning sulphur candles inside empty wine vessels kept them fresh and free from vinegar smell.

Structure and Bonding

SO_2 is a bent molecule with C_{2v} symmetry point group. A valence bond theory approach which just considered s and p orbitals would describe the bonding in terms of resonance between two resonance structures.

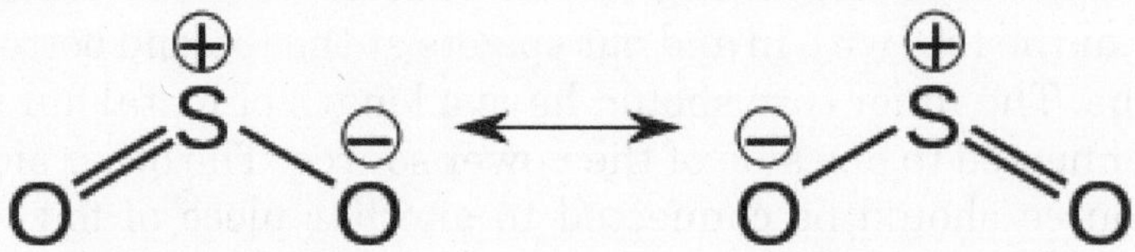

Figure: Two resonance structures of sulphur dioxide

The sulphur - oxygen bond has a bond order of 1.5. There is support for this simple approach that does not invoke d orbital participation. In terms of electron-counting formalism, the sulphur atom has an oxidation state of +4 and a formal charge of +1.

Occurrence

It is found on Earth and exists in very small concentrations and in the atmosphere at about 1 ppb.

On other planets, it can be found in various concentrations, the most significant being the atmosphere of Venus, where it is the third-most significant atmospheric gas at 150 ppm. There, it condenses to form clouds, and is a key component of chemical reactions in the planet's atmosphere and contributes to global warming. It has been implicated as a key agent in the warming of early Mars, with estimates of concentrations in the lower atmosphere as high as 100 ppm, though it only exists in trace amounts. On both Venus and Mars, its primary source, like on Earth, is believed to be volcanic. It is also believed to exist in trace amounts in the atmosphere of Jupiter.

As an ice, it is thought to exist in abundance on the Galilean moons – as sublimating ice or frost on the trailing hemisphere of Io, a natural satellite of Jupiter and in the crust and mantle of Europa, Ganymede, and Callisto, possibly also in liquid form and readily reacting with water.

Production

Sulphur dioxide is primarily produced for sulphuric acid manufacture. In the United States in 1979, 23.6 million tonnes of sulphur dioxide were used in this way, compared with 150 thousand tonnes used for other purposes. Most sulphur dioxide is produced by the combustion of elemental sulphur. Some sulphur dioxide is also produced by roasting pyrite and other sulfide ores in air.

Combustion Routes

Sulphur dioxide is the product of the burning of sulphur or of burning materials that contain sulphur:

$$S + O_2 \rightarrow SO_2, \Delta H = -297 \text{ kJ/mol}$$

To aid combustion, liquefied sulphur (140–150°C) is sprayed through an atomizing nozzle to generate fine drops of sulphur with a large surface area.

The reaction is exothermic, and the combustion produces temperatures of 1000–1600°C. The significant amount of heat produced is recovered by steam generation that can subsequently be converted to electricity.

The combustion of hydrogen sulfide and organosulfur compounds proceeds similarly. For example:

$$2\ H_2S + 3\ O_2 \rightarrow 2\ H_2O + 2\ SO_2$$

The roasting of sulfide ores such as pyrite, sphalerite, and cinnabar (mercury sulfide) also releases SO_2:

$$4\ FeS_2 + 11\ O_2 \rightarrow 2\ Fe_2O_3 + 8\ SO_2$$

$$2\ ZnS + 3\ O_2 \rightarrow 2\ ZnO + 2\ SO_2$$

$$HgS + O_2 \rightarrow Hg + SO_2$$

$$4\ FeS + 7O_2 \rightarrow 2\ Fe_2O_3 + 4\ SO_2$$

A combination of these reactions is responsible for the largest source of sulphur dioxide, volcanic eruptions. These events can release millions of tonnes of SO_2.

Reduction of Higher Oxides

Sulphur dioxide can also be a byproduct in the manufacture of calcium silicate cement; $CaSO_4$ is heated with coke and sand in this process:

$$2\ CaSO_4 + 2\ SiO_2 + C \rightarrow 2\ CaSiO_3 + 2\ SO_2 + CO_2$$

Until the 1970s, commercial quantities of sulphuric acid and cement were produced by this process in Whitehaven, England. Upon being mixed with shale or marl, and roasted, the sulphate liberated sulphur dioxide gas, used in sulphuric acid production, the reaction also produced calcium silicate, a precursor in cement production.

On a laboratory scale, the action of hot sulphuric acid on copper turnings produces sulphur dioxide.

$$Cu + 2\ H_2SO_4 \rightarrow CuSO_4 + SO_2 + 2\ H_2O$$

From Sulfite

Sulphite results from the reaction of aqueous base and sulphur dioxide. The reverse reaction involves acidification of sodium metabisulfite:

$$H_2SO_4 + Na_2S_2O_5 \rightarrow 2\ SO_2 + Na_2SO_4 + H_2O$$

Reactions

Industrial Reactions

Treatment of basic solutions with sulphur dioxide affords sulphite salts:

$$SO_2 + 2\ NaOH \rightarrow Na_2SO_3 + H_2O$$

Featuring sulphur in the +4 oxidation state, sulphur dioxide is a reducing agent. It is oxidized by halogens to give the sulphuryl halides, such as sulphuryl chloride:

$$SO_2 + Cl_2 \rightarrow SO_2Cl_2$$

Sulphur dioxide is the oxidising agent in the Claus process, which is conducted on a large scale in oil refineries. Here, sulphur dioxide is reduced by hydrogen sulfide to give elemental sulphur:

$$SO_2 + 2\ H_2S \rightarrow 3\ S + 2\ H_2O$$

The sequential oxidation of sulphur dioxide followed by its hydration is used in the production of sulphuric acid.

$$2\ SO_2 + 2\ H_2O + O_2 \rightarrow 2\ H_2SO_4$$

Laboratory Reactions

Sulphur dioxide is one of the few common acidic yet reducing gases. It turns moist litmus pink (being acidic), then white (due to its bleaching effect). It may be identified by bubbling it through a dichromate solution, turning the solution from orange to green (Cr^{3+} (aq)). It can also reduce ferric ions to ferrous

Sulphur dioxide can react with certain 1,3-dienes in a cheletropic reaction to form cyclic sulfones. This reaction is exploited on an industrial scale for the synthesis of sulfolane, which is an important solvent in the petrochemical industry.

Sulphur dioxide can bind to metal ions as a ligand to form metal sulphur dioxide complexes, typically where the transition metal is in oxidation state 0 or +1. Many different bonding modes (geometries) are recognized, but in most cases, the ligand is monodentate, attached to the metal through sulphur, which can be either planar and pyramidal η^1.

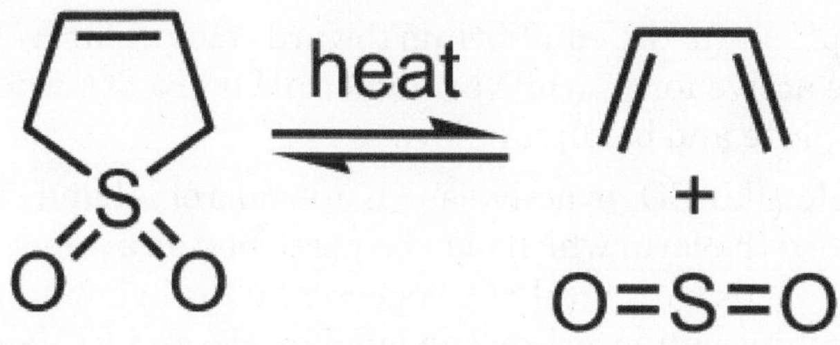

Uses

Precursor to Sulphuric Acid

Sulphur dioxide is an intermediate in the production of sulphuric acid, being converted to sulphur trioxide, and then to oleum, which is made into sulphuric acid. Sulphur dioxide for this purpose is made when sulphur combines with oxygen. The method of converting sulphur dioxide to sulphuric acid is called the contact process. Several billion kilograms are produced annually for this purpose.

As a Preservative

Sulphur dioxide is sometimes used as a preservative for dried apricots, dried figs, and other dried fruits, owing to its antimicrobial properties, and is called E220. when used in this way in Europe. As a preservative, it maintains the colourful appearance of the fruit and prevents rotting. It is also added to sulphured molasses.

In Winemaking

Sulphur dioxide was used by the Romans in winemaking, when they discovered that burning sulphur candles inside empty wine vessels keeps them fresh and free from vinegar smell.

Sulphur dioxide is still an important compound in winemaking, and is measured in parts per million in wine It is present even in so-called unsulphurated wine at concentrations of up to 10 mg/L. It serves as an antibiotic and antioxidant, protecting wine from spoilage by bacteria and oxidation. Its antimicrobial action also helps to minimize volatile acidity. Sulphur dioxide is responsible for the words "contains sulphites" found on wine labels.

Sulphur dioxide exists in wine in free and bound forms, and the combinations are referred to as total SO_2. Binding, for instance to the carbonyl group of acetaldehyde, varies with the wine in question. The free form exists in equilibrium between molecular SO_2 (as a dissolved gas) and bisulphite ion, which is in turn in equilibrium with sulphite ion. These equilibria depend on the pH of the wine.

Lower pH shifts the equilibrium towards molecular (gaseous) SO_2, which is the active form, while at higher pH more SO_2 is found in the inactive sulphite and bisulphite forms.

The molecular SO_2 is active as an antimicrobial and antioxidant, and this is also the form which may be perceived as a pungent odour at high levels. Wines with total SO_2 concentrations below 10 ppm do not require "contains sulphites" on the label by US and EU laws.

The upper limit of total SO_2 allowed in wine in the US is 350 ppm; in the EU it is 160 ppm for red wines and 210 ppm for white and rosé wines. In low concentrations, SO_2 is mostly undetectable in wine, but at free SO_2 concentrations over 50 ppm, SO_2 becomes evident in the nose and taste of wine.

SO_2 is also a very important compound in winery sanitation. Wineries and equipment must be kept clean, and because bleach cannot be used in a winery due the risk of cork taint, a mixture of SO_2, water, and citric acid is commonly used to clean and sanitize equipment. Compounds of ozone (O_3) are now used extensively as cleaning products in wineries due to their efficiency, and because these compounds do not affect the wine or equipment.

As a Reducing Agent

Sulphur dioxide is also a good reductant. In the presence of water, sulphur dioxide is able to decolorize substances. Specifically, it is a useful reducing bleach for papers and delicate materials such as clothes. This bleaching effect normally does not last very long.

Oxygen in the atmosphere reoxidizes the reduced dyes, restoring the colour. In municipal wastewater treatment, sulphur dioxide is used to treat chlorinated wastewater prior to release. Sulphur dioxide reduces free and combined chlorine to chloride.

Sulphur dioxide is fairly soluble in water, and by both IR and Raman spectroscopy; the hypothetical sulphurous acid, H_2SO_3, is not present to any extent. However, such solutions do show spectra of the hydrogen sulphite ion, HSO_3^-, by reaction with water, and it is in fact the actual reducing agent present:

$$SO_2 + H_2O \rightleftharpoons HSO_3^- + H^+$$

Biochemical and Biomedical Roles

Sulphur dioxide is toxic in large amounts. It or its conjugate base bisulphite is produced biologically as an intermediate in both sulphate-reducing organisms and in sulphur-oxidizing bacteria, as well. The role of sulphur dioxide in mammalian biology is not yet well understood. Sulphur dioxide blocks nerve signals from the pulmonary stretch receptors and abolishes the Hering–Breuer inflation reflex.

As a Refrigerant

Being easily condensed and possessing a high heat of evaporation, sulphur dioxide is a candidate material for refrigerants. Prior to the development of chlorofluorocarbons, sulphur dioxide was used as a refrigerant in home refrigerators.

As a Reagent and Solvent in the Laboratory

Sulphur dioxide is a versatile inert solvent widely used for dissolving highly oxidizing salts. It is also used occasionally as a source of the sulphonyl group in organic synthesis. Treatment of aryl diazonium salts with sulphur dioxide and cuprous chloride yields the corresponding aryl sulphonyl chloride, for example:

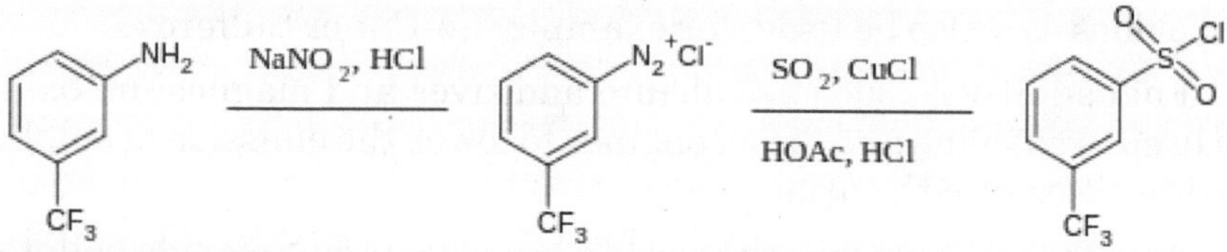

As an Air Pollutant

Sulphur dioxide is a noticeable component in the atmosphere, especially following volcanic eruptions. According to the United States Environmental Protection Agency, the amount of sulphur dioxide released in the U.S. per year was:

Year	SO_2
1970	31,161,000 short tons (28.3 Mt)
1980	25,905,000 short tons (23.5 Mt)
1990	23,678,000 short tons (21.5 Mt)
1996	18,859,000 short tons (17.1 Mt)
1997	19,363,000 short tons (17.6 Mt)
1998	19,491,000 short tons (17.7 Mt)
1999	18,867,000 short tons (17.1 Mt)

Sulphur dioxide is a major air pollutant and has significant impacts upon human health. In addition, the concentration of sulphur dioxide in the atmosphere can influence the habitat suitability for plant communities, as well as animal life. Sulphur dioxide emissions are a precursor to acid rain and atmospheric particulates. Due largely to the US EPA's Acid Rain Programme, the U.S. has had a 33% decrease in emissions between 1983 and 2002. This improvement resulted in part from flue-gas desulfurization, a technology that enables SO_2 to be chemically bound in power plants burning sulphur-containing coal or oil. In particular, calcium oxide (lime) reacts with sulphur dioxide to form calcium sulphite:

$$CaO + SO_2 \rightarrow CaSO_3$$

Aerobic oxidation of the $CaSO_3$ gives $CaSO_4$, anhydrite. Most gypsum sold in Europe comes from flue-gas desulfurization.

Sulphur can be removed from coal during the burning process by using limestone as a bed material in fluidized bed combustion.

Sulphur can also be removed from fuels prior to burning the fuel, preventing the formation of SO_2 because no sulphur remains in the fuel from which SO_2 can be formed. The Claus process is used in refineries to produce sulphur as a byproduct. The Stretford process has also been used to remove sulphur from fuel. Redox processes using iron oxides can also be used, for example, Lo-Cat or Sulferox.

Fuel additives, such as calcium additives and magnesium oxide, are used in gasoline and diesel engines to lower the emission of sulphur dioxide gases into the atmosphere.

As of 2006, China was the world's largest sulphur dioxide polluter, with 2005 emissions estimated to be 25,490,000 short tons (23.1 Mt). This amount represents a 27% increase since 2000, and is roughly comparable with U.S. emissions in 1980.

Safety

Inhalation

Inhaling sulphur dioxide is associated with increased respiratory symptoms and disease, difficulty in breathing, and premature death. In 2008, the American Conference of Governmental Industrial Hygienists reduced the short-term exposure limit to 5 parts per million (ppm). The OSHA PEL is currently set at 2 ppm (5.64 mg/m^3) time-weighted average. NIOSH has set the IDLH at 100 ppm. In 2010, the EPA "revised the primary SO_2 NAAQS by establishing a new one-hour standard at a level of 75 parts per billion (ppb). EPA revoked the two existing primary standards because they would not provide additional public health protection given a one-hour standard at 75 ppb."

A 2011 systematic review concluded that exposure to sulphur dioxide is associated with preterm birth.

Ingestion

In the United States, the Centre for Science in the Public Interest lists the two food preservatives, sulphur dioxide and sodium bisulphite, as being safe for human consumption except for certain asthmatic individuals who may be sensitive to them, especially in large amounts. Symptoms of sensitivity to sulphiting agents, including sulphur dioxide, manifest as potentially life-threatening trouble breathing within minutes of ingestion.

Nitrogen Dioxide

Nitrogen dioxide is the chemical compound with the formula NO_2. It is one of several nitrogen oxides. NO_2 is an intermediate in the industrial synthesis of nitric acid, millions of tons of which are produced each year. This reddish-brown toxic gas has a characteristic sharp, biting odor and is a prominent air pollutant. Nitrogen dioxide is a paramagnetic, bent molecule with C_{2v} point group symmetry.

Molecular Properties

Nitrogen dioxide has a molar mass of 46.0055, which makes it heavier than air, whose average molar mass is 28.8. The bond length between the nitrogen atom and the oxygen atom is 119.7 pm. This bond length is consistent with a bond order between one and two.

Unlike ozone, O_3, the ground electronic state of nitrogen dioxide is a doublet state, since nitrogen has one unpaired electron, which decreases the alpha effect compared to nitrite and creates a weak bonding interaction with the oxygen lone pairs. The lone electron in NO_2 also means that this compound is a free radical, so the formula for nitrogen dioxide is often written as $\cdot NO_2$.

Preparation and Reactions

Nitrogen dioxide typically arises via the oxidation of nitric oxide by oxygen in air:

$$2\,NO + O_2 \rightarrow 2\,NO_2$$

In the laboratory, NO_2 can be prepared in a two-step procedure where dehydration of nitric acid produces dinitrogen pentoxide, which subsequently undergoes thermal decomposition:

$$2\,HNO_3 \rightarrow N_2O_5 + H_2O$$

$$2\,N_2O_5 \rightarrow 4\,NO_2 + O_2$$

The thermal decomposition of some metal nitrates also affords NO_2:

$$2\ Pb(NO_3)_2 \rightarrow 2\ PbO + 4\ NO2 + O2$$

Alternatively, reduction of concentrated nitric acid by metal (such as copper).

$$4\ HNO3 + Cu \rightarrow Cu(NO_3)_2 + 2\ NO2 + 2\ H_2O$$

Or finally by adding concentrated nitric acid over tin; hydrated tin dioxide is produced as byproduct.

$$4HNO_3 + Sn \rightarrow H_2O + H_2SnO_3 + 4\ NO_2$$

Main Reactions

Basic Thermal Properties

NO_2 exists in equilibrium with the colourless gas dinitrogen tetroxide (N_2O4):

$$2\ NO_2 \rightleftharpoons N_2O_4$$

The equilibrium is characterized by ΔH = –57.23 kJ/mol, which is exothermic. NO_2 is favoured at higher temperatures, while at lower temperatures, dinitrogen tetroxide (N_2O_4) predominates. Dinitrogen tetroxide (N_2O_4) can be obtained as a white solid with melting point –11.2 °C. NO_2 is paramagnetic due to its unpaired electron, while N_2O_4 is diamagnetic.

The chemistry of nitrogen dioxide has been investigated extensively. At 150 °C, NO_2 decomposes with release of oxygen via an endothermic process (ΔH = 114 kJ/mol):

$$2\ NO_2 \rightarrow 2\ NO + O_2$$

As an Oxidizer

As suggested by the weakness of the N–O bond, NO_2 is a good oxidizer. Consequently, it will combust, sometimes explosively, with many compounds, such as hydrocarbons.

Hydrolysis

It hydrolyses to give nitric acid and nitrous acid:

$$2\ NO_2/N_2O_4 + H_2O \rightarrow HNO_2 + HNO_3$$

This reaction is one step in the Ostwald process for the industrial production of nitric acid from ammonia. Nitric acid decomposes slowly to nitrogen dioxide, which confers the characteristic yellow colour of most samples of this acid:

$$4\ HNO_3 \rightarrow 4\ NO_2 + 2\ H_2O + O_2$$

Conversion to Nitrates

NO_2 is used to generate anhydrous metal nitrates from the oxides:

$$MO + 3\ NO_2 \rightarrow M(NO_3)_2 + NO$$

Alkyl and metal iodides give the corresponding nitrites:

$$2\ CH_3I + 2\ NO_2 \rightarrow 2\ CH_3NO_2 + I_2$$

$$TiI_4 + 4\ NO_2 \rightarrow Ti(NO_2)4 + 2\ I_2$$

Safety and Pollution Considerations

Nitrogen dioxide is toxic by inhalation. However, as the compound is acrid and easily detectable by smell at low concentrations, inhalation exposure can generally be avoided.

One potential source of exposure is white fuming nitric acid, which spontaneously produces NO_2 above 0 °C. Symptoms of poisoning (lung edema) tend to appear several hours after inhalation of a low but potentially fatal dose. Also, low concentrations (4 ppm) will anesthetize the nose, thus creating a potential for overexposure.

There is some evidence that long-term exposure to NO_2 at concentrations above 40–100 μg/m^3 may decrease lung function and increase the risk of respiratory symptoms.

Nitrogen dioxide is formed in most combustion processes using air as the oxidant. At elevated temperatures nitrogen combines with oxygen to form nitric oxide:

$$O_2 + N_2 \rightarrow 2\ NO$$

Nitric oxide can be oxidized in air to form nitrogen dioxide. At normal atmospheric concentrations, this is a very slow process.

$$2\ NO + O_2 \rightarrow 2\ NO_2$$

The most prominent sources of NO_2 are internal combustion engines, thermal power stations and, to a lesser extent, pulp mills. Butane gas heaters and stoves are also sources. The excess air required for complete combustion of fuels in these processes introduces nitrogen into the combustion reactions at high temperatures and produces nitrogen oxides (NO*x*). Limiting NO*x* production demands the precise control of the amount of air used in combustion. In households, kerosene heaters and gas heaters are sources of nitrogen dioxide.

Nitrogen dioxide is also produced by atmospheric nuclear tests, and is responsible for the reddish colour of mushroom clouds.

Nitrogen dioxide is a large scale pollutant, with rural background ground level concentrations in some areas around 30 μg/m^3, not far below unhealthy levels. Nitrogen dioxide plays a role in atmospheric chemistry, including the formation of tropospheric ozone. A 2005 study by researchers at the University of California, San Diego, suggests a link between NO_2 levels and Sudden Infant Death Syndrome.

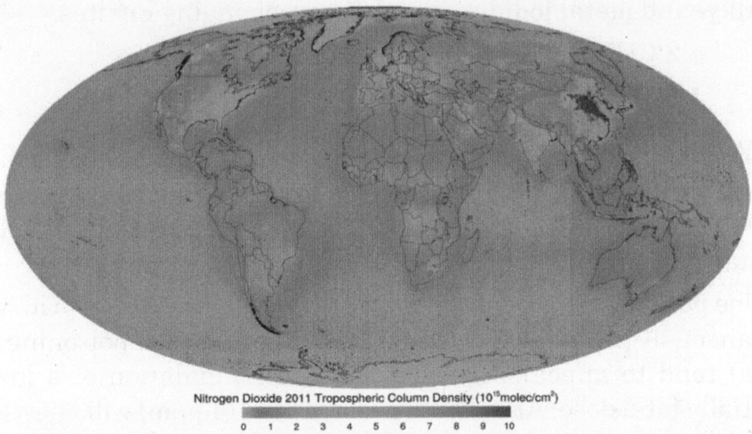

Figure: *Nitrogen Dioxide 2011 tropospheric column density.*

Nitrogen dioxide is also produced naturally during electrical storms. The term for this process is "atmospheric fixation of nitrogen". The rain produced during such storms is especially good for the garden as it contains trace amounts of fertilizer. (Henry Cavendish 1784, Birkland -Eyde Process 1903, et-al)

Ammonia

Ammonia, or azane, is a compound of nitrogen and hydrogen with the formula NH_3. It is a colourless gas with a characteristic pungent smell. Ammonia contributes significantly to the nutritional needs of terrestrial organisms by serving as a precursor to food and fertilizers. Ammonia, either directly or indirectly, is also a building-block for the synthesis of many pharmaceuticals and is used in many commercial cleaning products. Although in wide use, ammonia is both caustic and hazardous. The global industrial production of ammonia for 2012 was anticipated to be 198,000,000 tonnes (195,000,000 long tons; 218,000,000 short tons) , a 35% increase over the estimated 2006 global output of 146,500,000 tonnes (144,200,000 long tons; 161,500,000 short tons) .

NH_3 boils at –33.34 °C (–28.012 °F) at a pressure of one atmosphere, so the liquid must be stored under pressure or at low temperature. Household ammonia or ammonium hydroxide is a solution of NH_3 in water. The concentration of such solutions is measured in units of the Baumé scale (density), with 26 degrees baumé (about 30% (by weight) ammonia at 15.5 °C or 59.9 °F) being the typical high-concentration commercial product.

Natural Occurrence

Ammonia is found in trace quantities in the atmosphere, being produced from the putrefaction (decay process) of nitrogenous animal and vegetable matter. Ammonia and ammonium salts are also found in small quantities in rainwater, whereas ammonium chloride (sal-ammoniac), and ammonium sulphate are found in volcanic districts; crystals of ammonium bicarbonate have been found in Patagonian guano. The kidneys secrete ammonia to neutralize excess acid. Ammonium salts are found distributed through fertile soil and in seawater. Ammonia is also found throughout the Solar System on Pluto, Mars, Jupiter, Saturn, Uranus, and Neptune. Substances containing ammonia, or those that are similar to it, are called *ammoniacal.*

Properties

Ammonia is a colourless gas with a characteristic pungent smell. It is lighter than air, its density being 0.589 times that of air. It is easily liquefied due to the strong hydrogen bonding between molecules; the liquid boils at –33.3 °C (–27.94 °F), and freezes at –77.7 °C (–107.86 °F) to white crystals. Ammonia may be conveniently deodorized by reacting it with either sodium bicarbonate or acetic acid. Both of these reactions form an odourless ammonium salt.

Solid

The crystal symmetry is cubic, Pearson symbol cP16, space group $P2_13$ No.198, lattice constant 0.5125 nm.

Liquid

Liquid ammonia possesses strong ionising powers reflecting its high ε of 22. Liquid ammonia has a very high standard enthalpy change of vaporization (23.35 kJ/mol, *cf.* water 40.65 kJ/mol, methane 8.19 kJ/mol, phosphine 14.6 kJ/mol) and can therefore be used in laboratories in uninsulated vessels without additional refrigeration.

Solvent Properties

Ammonia is miscible with water. In an aqueous solution, it can be expelled by boiling. The aqueous solution of ammonia is basic. The maximum concentration of ammonia in water (a saturated solution) has a density of 0.880 g/cm^3 and is often known as '.880 ammonia'. Ammonia does not burn readily or sustain combustion, except under narrow fuel-to-air mixtures of 15–25% air.

Combustion

When mixed with oxygen, it burns with a pale yellowish-green flame. At high temperature and in the presence of a suitable catalyst, ammonia is decomposed into its constituent elements. Ignition occurs when chlorine is passed into ammonia, forming nitrogen and hydrogen chloride; if chlorine is present in excess, then the highly explosive nitrogen trichloride (NCl_3) is also formed.

Structure

The ammonia molecule has a trigonal pyramidal shape as predicted by the valence shell electron pair repulsion theory (VSEPR theory) with an experimentally determined bond angle of 106.7°. The central nitrogen atom has five outer electrons with an additional electron from each hydrogen atom. This gives a total of eight electrons, or four electron pairs that are arranged tetrahedrally. Three of these electron pairs are used as bond pairs, which leaves one lone pair of electrons. The lone pair of electrons repel more strongly than bond pairs, therefore the bond angle is not 109.5°, as expected for a regular tetrahedral arrangement, but is measured at 106.7°.

The nitrogen atom in the molecule has a lone electron pair, which makes ammonia a base, a proton acceptor. This shape gives the molecule a dipole moment and makes it polar. The molecule's polarity and, especially, its ability to form hydrogen bonds, makes ammonia highly miscible with water. Ammonia is moderately basic, a 1.0 M aqueous solution has a pH of 11.6 and if a strong acid is added to such a solution until the solution is neutral (pH = 7), 99.4% of the ammonia molecules are protonated. Temperature and salinity also affect the proportion of NH_4^+. The latter has the shape of a regular tetrahedron and is isoelectronic with methane.

The ammonia molecule readily undergoes nitrogen inversion at room temperature; a useful analogy is an umbrella turning itself inside out in a strong wind. The energy barrier to this inversion is 24.7 kJ/mol, and the resonance frequency is 23.79 GHz, corresponding to microwave radiation of a wavelength of 1.260 cm. The absorption at this frequency was the first microwave spectrum to be observed.

Amphotericity

One of the most characteristic properties of ammonia is its basicity. Ammonia is considered to be a weak base. It combines with acids to form salts; thus with hydrochloric acid it forms ammonium chloride (sal-ammoniac); with nitric acid, ammonium nitrate, etc. However, perfectly dry ammonia will not combine with perfectly dry hydrogen chloride; moisture is necessary to bring about the reaction. As a demonstration experiment, opened bottles of concentrated ammonia and hydrochloric acid produce clouds of ammonium chloride, which seem to appear "out of nothing" as the salt forms where the two diffusing clouds of molecules meet, somewhere between the two bottles.

$$NH_3 + HCl \rightarrow NH_4Cl$$

The salts produced by the action of ammonia on acids are known as the ammonium salts and all contain the ammonium ion (NH_4^+).

Although ammonia is well known as a weak base, it can also act as an extremely weak acid. It is a protic substance and is capable of formation of amides (which contain the NH_2^- ion). For example, lithium dissolves in liquid ammonia to give a solution of lithium amide:

$$Li + NH_3 \rightarrow LiNH_2 + ½\ H_2$$

Self-dissociation

Like water, ammonia undergoes molecular autoionisation to form its acid and base conjugates:

$$2\ NH_3\ (aq) \rightleftharpoons NH^+_4\ (aq) + NH^-_2\ (aq)$$

At standard pressure and temperature, $K=[NH+4][NH^-_2] = 10^-_{30}$

Combustion

The combustion of ammonia to nitrogen and water is exothermic:

$$4\ NH_3 + 3\ O_2 \rightarrow 2\ N_2 + 6\ H_2O\ (g)\ (\Delta H^\circ_r = -1267.20\ kJ/mol)$$

The standard enthalpy change of combustion, ΔH°_c, expressed per mole of ammonia and with condensation of the water formed, is –382.81 kJ/mol. Dinitrogen is the thermodynamic product of combustion: all nitrogen oxides are unstable with respect to N_2 and O_2, which is the principle behind the catalytic converter. However, nitrogen oxides can be formed as kinetic products in the presence of appropriate catalysts, a reaction of great industrial importance in the production of nitric acid:

$$4\ NH_3 + 5\ O_2 \rightarrow 4\ NO + 6\ H_2O$$

A subsequent reaction leads to NO_2

$$2\ NO + O_2 \rightarrow 2\ NO_2$$

The combustion of ammonia in air is very difficult in the absence of a catalyst (such as platinum gauze), as the temperature of the flame is usually lower than the ignition temperature of the ammonia-air mixture. The flammable range of ammonia in air is 16–25%.

Formation of Other Compounds

In organic chemistry, ammonia can act as a nucleophile in substitution reactions. Amines can be formed by the reaction of ammonia with alkyl halides, although the resulting $-NH_2$ group is also nucleophilic and secondary and tertiary amines are often formed as byproducts. An excess of ammonia helps minimise multiple substitution, and neutralises the hydrogen halide formed. Methylamine is prepared commercially by the reaction of ammonia with chloromethane, and the reaction of ammonia with 2-bromopropanoic acid has been used to prepare racemic alanine in 70% yield. Ethanolamine is prepared by a ring-opening reaction with ethylene oxide: the reaction is sometimes allowed to go further to produce diethanolamine and triethanolamine.

Amides can be prepared by the reaction of ammonia with a number of carboxylic acid derivatives. Acyl chlorides are the most reactive, but the ammonia must be present in at least a twofold excess to neutralise the hydrogen chloride formed. Esters and anhydrides also react with ammonia to form amides. Ammonium salts of carboxylic acids can be dehydrated to amides so long as there are no thermally sensitive groups present: temperatures of 150–200 °C are required.

The hydrogen in ammonia is capable of replacement by metals, thus magnesium burns in the gas with the formation of magnesium nitride Mg_3N_2, and when the gas is passed over heated sodium or potassium, sodamide, $NaNH_2$, and potassamide, KNH_2, are formed. Where necessary in substitutive nomenclature, IUPAC recommendations prefer the name “azane” to ammonia: hence chloramine would be named “chloroazane” in substitutive nomenclature, not “chloroammonia”.

Pentavalent ammonia is known as λ^5-amine, or more commonly, ammonium hydride. This crystalline solid is only stable under high pressure, and decomposes back into trivalent ammonia and hydrogen gas at normal conditions. This substance was once investigated as a possible solid rocket fuel in 1966.

Ammonia as a Ligand

Ammonia can act as a ligand in transition metal complexes. It is a pure σ-donor, in the middle of the spectrochemical series, and shows intermediate hard-soft behaviour. For historical reasons, ammonia is named ammine in the nomenclature of coordination compounds. Some notable ammine complexes include tetraamminediaquacopper(II) ($[Cu(NH_3)_4(H_2O)_2]^{2+}$), a dark blue complex formed by adding ammonia to solution of copper(II) salts. It is known as Schweizer’s reagent.

Diamminesilver(I) ($[Ag(NH_3)_2]^+$) is the active species in Tollens' reagent. Formation of this complex can also help to distinguish between precipitates of the different silver halides: silver chloride (AgCl) is soluble in dilute (2M) ammonia solution, silver bromide (AgBr) is only soluble in concentrated ammonia solution, whereas silver iodide (AgI) is insoluble in aqueous ammonia.

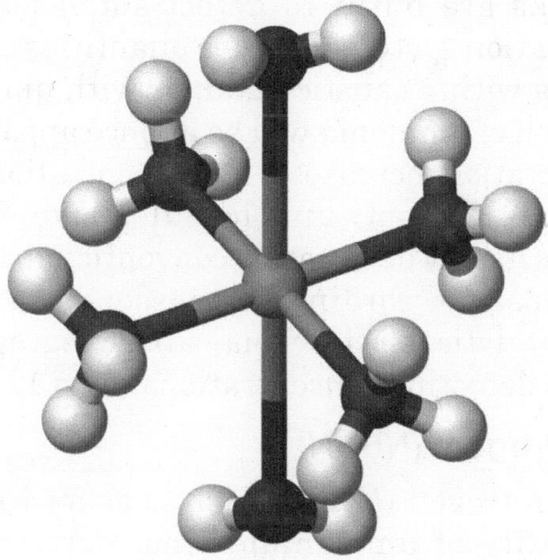

***Figure:** Ball-and-stick model of the tetraamminediaquacopper(II) cation, $[Cu(NH_3)_4(H_2O)_2]^{2+}$*

Ammine complexes of chromium(III) were known in the late 19th century, and formed the basis of Alfred Werner's revolutionary theory on the structure of coordination compounds. Werner noted only two isomers (*fac*- and *mer*-) of the complex $[CrCl_3(NH_3)_3]$ could be formed, and concluded the ligands must be arranged around the metal ion at the vertices of an octahedron. This proposal has since been confirmed by X-ray crystallography.

An ammine ligand bound to a metal ion is markedly more acidic than a free ammonia molecule, although deprotonation in aqueous solution is still rare. One example is the Calomel reaction, where the resulting amidomercury(II) compound is highly insoluble.

$$Hg_2Cl_2 + 2\ NH_3 \rightarrow Hg + HgCl(NH_2) + NH_4^+ + Cl^-$$

Ammonia in Solution

Ammonia and ammonium salts can be readily detected, in very minute traces, by the addition of Nessler's solution, which gives a distinct yellow colouration in the presence of the least trace of ammonia or ammonium salts. The amount of ammonia in ammonium salts can be estimated quantitatively by distillation of the salts with sodium or potassium hydroxide, the ammonia evolved being absorbed in a known volume of standard sulphuric acid and the excess of acid then determined volumetrically; or the ammonia may be absorbed in hydrochloric acid and the ammonium chloride so formed precipitated as ammonium hexachloroplatinate, $(NH_4)_2PtCl_6$.

Gaseous Ammonia

Sulphur sticks are burnt to detect small leaks in industrial ammonia refrigeration systems. Larger quantities can be detected by warming the salts with a caustic alkali or with quicklime, when the characteristic smell of ammonia will be at once apparent. Ammonia is an irritant and irritation increases with concentration; the Permissible Exposure Limit is 25ppm, and lethal above 500ppm. Higher concentrations are hardly detected by conventional detectors, the type of detector is chosen according to the sensitivity required (e.g. semiconductor, catalytic, electrochemical). Holographic sensors have been proposed for detecting concentrations up to 12.5% in volume.

Ammoniacal Nitrogen (NH_3-N)

Ammoniacal nitrogen (NH_3-N) is a measure commonly used for testing the quantity of ammonium ions, derived naturally from ammonia, and returned to ammonia via organic processes, in water or waste liquids. It is a measure used mainly for quantifying values in waste treatment and water purification systems, as well as a measure of the health of natural and man made water reserves. It is measured in units of mg/L (milligram per litre).

Uses

Fertilizer

Approximately 83% (as of 2004) of ammonia is used as fertilizers either as its salts, solutions or anhydrously. When applied to soil, it helps provide increased yields of crops such as maize and wheat. 30% of agricultural nitrogen applied in the USA is in the form of anhydrous ammonia and worldwide 110 million tonnes are applied each year.

Precursor to Nitrogenous Compounds

Ammonia is directly or indirectly the precursor to most nitrogen-containing compounds. Virtually all synthetic nitrogen compounds are derived from ammonia. An important derivative is nitric acid. This key material is generated via the Ostwald process by oxidation of ammonia with air over a platinum catalyst at 700–850 °C (1,292–1,562 °F), ~9 atm. Nitric oxide is an intermediate in this conversion:

$$NH_3 + 2\,O_2 \rightarrow HNO_3 + H_2O$$

Nitric acid is used for the production of fertilizers, explosives, and many organonitrogen compounds.

Ammonia is also used to make the following compounds:

- Hydrazine, in the Olin Raschig process, the Ketazine process, and the peroxide process
- Hydrogen cyanide, in the BMA process and the Andrussow process
- Hydroxylamine and ammonium carbonate, in the Raschig process
- Phenol, in the Raschig–Hooker process
- Urea, in the Bosch–Meiser urea process and in Wöhler synthesis
- Amino acids, using Strecker amino-acid synthesis
- Acrylonitrile, in the Sohio process

Ammonia can also be used to make compounds in reactions which are not specifically named. Examples of such compounds include: ammonium perchlorate, ammonium nitrate, formamide, dinitrogen tetroxide, alprazolam, ethanolamine, ethyl carbamate, hexamethylenetetramine, and ammonium bicarbonate.

Cleaner

Household ammonia is a solution of NH_3 in water (i.e., ammonium hydroxide) used as a general purpose cleaner for many surfaces. Because ammonia results in a relatively streak-free shine, one of its most common uses is to clean glass, porcelain and stainless steel. It is also frequently used for cleaning ovens and soaking items to loosen baked-on grime. Household ammonia ranges in concentration by weight from 5 to 10% ammonia.

Fermentation

Solutions of ammonia ranging from 16% to 25% are used in the fermentation industry as a source of nitrogen for microorganisms and to adjust pH during fermentation.

Antimicrobial Agent for Food Products

As early as in 1895, it was known that ammonia was "strongly antiseptic ... it requires 1.4 grams per litre to preserve beef tea." In one study, anhydrous ammonia destroyed 99.999% of zoonotic bacteria in 3 types of animal feed, but not silage. Anhydrous ammonia is currently used commercially to reduce or eliminate microbial contamination of beef. Pink slime (or lean finely textured beef in the beef industry) is made from fatty beef trimmings (about 50–70% fat) by removing the fat using heat and centrifugation, then treating it with ammonia to kill *E. coli*. The process was deemed effective and safe by the US Department of Agriculture based on a study (funded by a producer of pink slime) that found that the treatment reduces *E. coli* to undetectable levels. There have been safety concerns about the process as well as consumer complaints about the taste and smell of beef treated at optimal levels of ammonia. However, the level of ammonia in any final product has not come close to toxic levels to humans.

Minor and Emerging Uses

Refrigeration – R717

Because of ammonia's vaporization properties, it is a useful refrigerant. It was commonly used prior to the popularisation of chlorofluorocarbons (Freons). Anhydrous ammonia is widely used in industrial refrigeration applications and hockey rinks because of its high energy efficiency and low cost. However, it suffers from the disadvantage of toxicity, which restricts its domestic and small-scale use. Along with its use in modern vapour-compression refrigeration it was used in a mixture along with hydrogen and water in absorption refrigerators. The Kalina cycle, which is of growing importance to geothermal power plants, depends on the wide boiling range of the ammonia-water mixture.

For Remediation of Gaseous Emissions

Ammonia is used to scrub SO_2 from the burning of fossil fuels, and the resulting product is converted to ammonium sulphate for use as fertilizer. Ammonia neutralizes the nitrogen oxides (NO_x) pollutants emitted by diesel engines. This technology, called SCR (selective catalytic reduction), relies on a vanadia-based catalyst. Ammonia may be used to mitigate gaseous spills of phosgene.

Carbon Monoxide

Carbon monoxide (CO) is a colourless, odourless, and tasteless gas that is slightly less dense than air. It is toxic to humans when encountered in concentrations above about 35 ppm, although it is also produced in normal animal metabolism in low quantities, and is thought to have some normal biological functions. In the atmosphere, it is spatially variable and short lived, having a role in the formation of ground-level ozone.

Carbon monoxide consists of one carbon atom and one oxygen atom, connected by a triple bond that consists of two covalent bonds as well as one dative covalent bond. It is the simplest oxocarbon, and isoelectronic with the cyanide ion and molecular nitrogen. In coordination complexes the carbon monoxide ligand is called carbonyl.

Carbon monoxide is produced from the partial oxidation of carbon-containing compounds; it forms when there is not enough oxygen to produce carbon dioxide (CO_2), such as when operating a stove or an internal combustion engine in an enclosed space. In the presence of oxygen, including atmospheric concentrations, carbon monoxide burns with a blue flame, producing carbon dioxide. Coal gas, which was widely used before the 1960s for domestic lighting, cooking, and heating, had carbon monoxide as a significant fuel constituent. Some processes in modern technology, such as iron smelting, still produce carbon monoxide as a byproduct.

Worldwide, the largest source of carbon monoxide is natural in origin, due to photochemical reactions in the troposphere that generate about 5×10^{12} kilograms per year. Other natural sources of CO include volcanoes, forest fires, and other forms of combustion.

In biology, carbon monoxide is naturally produced by the action of heme oxygenase 1 and 2 on the heme from hemoglobin breakdown. This process produces a certain amount of carboxyhemoglobin in normal persons, even if they do not breathe any carbon monoxide. Following the first report that carbon monoxide is a normal neurotransmitter in 1993, as well as one of three gases that naturally modulate inflammatory responses in the body (the other two being nitric oxide and hydrogen sulfide), carbon monoxide has received a great deal of clinical attention as a biological regulator. In many tissues, all three gases are known to act as anti-inflammatories, vasodilators, and promoters of neovascular growth. Clinical trials of small amounts of carbon monoxide as a drug are ongoing.

History

Aristotle (384–322 BC) first recorded that burning coals emanated toxic fumes. An ancient method of execution was to shut the criminal in a bathing room with smouldering coals. What was not known was the mechanism of death. Greek physician Galen (129–199 AD) speculated that there was a change in the composition of the air that caused harm when inhaled. In 1776, the French chemist de Lassone produced CO by heating zinc oxide with coke, but mistakenly concluded that the gaseous product was hydrogen, as it burned with a blue flame. The gas was identified as a compound containing carbon and oxygen by the Scottish chemist William Cumberland Cruikshank in the year 1800. Its toxic properties on dogs were thoroughly investigated by Claude Bernard around 1846.

During World War II, a gas mixture including carbon monoxide was used to keep motor vehicles running in parts of the world where gasoline and diesel fuel were scarce. External (with a few exceptions) charcoal or wood gas generators were fitted, and the mixture of atmospheric nitrogen, carbon monoxide, and small amounts of other gases produced by gasification was piped to a gas mixer. The gas mixture produced by this process is known as wood gas. Carbon monoxide was also used on a large scale during the Holocaust at some Nazi extermination camps, the most notable by gas vans in Chelmno, and in the Action T4 "euthanasia" programme.

Molecular Properties

Carbon monoxide has a molar mass of 28.0, which makes it slightly lighter than air, whose average molar mass is 28.8. According to the ideal gas law, CO is therefore less dense than air.

The bond length between the carbon atom and the oxygen atom is 112.8 pm. This bond length is consistent with a triple bond, as in molecular nitrogen (N_2), which has a similar bond length and nearly the same molecular mass. Carbon–oxygen double bonds are significantly longer, 120.8 pm in formaldehyde, for example. The boiling point (82 K) and melting point (68 K) are very similar to those of N_2 (77 K and 63 K, respectively). The bond dissociation energy of 1072 kJ/mol is stronger than that of N_2 (942 kJ/mol) and represents the strongest chemical bond known.

The ground electronic state of carbon monoxide is a singlet state since there are no unpaired electrons.

Bonding and Dipole Moment

Carbon and oxygen together have a total of 10 valence electrons in carbon monoxide. To satisfy the octet rule for the carbon, the two atoms form a triple bond, with six shared electrons in three bonding molecular orbitals, rather than the usual double bond found in organic carbonyl compounds. Since four of the shared electrons come from the oxygen atom and only two from carbon, one bonding orbital is occupied by two electrons from oxygen, forming a dative or dipolar bond. This causes a $C \leftarrow O$ polarization of the molecule, with a small negative charge on carbon and a small positive charge on oxygen. The other two bonding orbitals are each occupied by one electron from carbon and one from oxygen, forming (polar) covalent bonds with a reverse $C \rightarrow O$ polarization, since oxygen is more electronegative than carbon. In the free carbon monoxide, a net negative charge δ- remains at the carbon end and the molecule has a small dipole moment of 0.122 D.

The molecule is therefore asymmetric: oxygen has more electron density than carbon, and is also slightly positively charged compared to carbon being negative. By contrast, the isoelectronic dinitrogen molecule has no dipole moment.

If carbon monoxide acts as a ligand, the polarity of the dipole may reverse with a net negative charge on the oxygen end, depending on the structure of the coordination complex.

Bond Polarity and Oxidation State

Theoretical and experimental studies show that, despite the greater electronegativity of oxygen, the dipole moment points from the more-negative carbon end to the more-positive oxygen end. The three bonds are in fact polar covalent bonds that are strongly polarized. The calculated polarization toward the oxygen atom is 71% for the σ-bond and 77% for both π-bonds.

The oxidation state of carbon in carbon monoxide is +2 in each of these structures. It is calculated by counting all the bonding electrons as belonging to the more electronegative oxygen. Only the two non-bonding electrons on carbon are assigned to carbon. In this count, carbon then has only two valence electrons in the molecule compared to four in the free atom.

Biological and Physiological Properties

Toxicity

Carbon monoxide poisoning is the most common type of fatal air poisoning in many countries. Carbon monoxide is colourless, odourless, and tasteless, but highly toxic. It combines with hemoglobin to produce carboxyhemoglobin, which usurps the space in hemoglobin that normally carries oxygen, but is ineffective for delivering oxygen to bodily tissues. Concentrations as low as 667 ppm may cause up to 50% of the body's hemoglobin to convert to carboxyhemoglobin. A level of 50% carboxyhemoglobin may result in seizure, coma, and fatality. In the United States, the OSHA limits long-term workplace exposure levels above 50 ppm. Within short time scales, carbon monoxide absorption is cumulative, since the half-life is about 5 hours in fresh air.

The most common symptoms of carbon monoxide poisoning may resemble other types of poisonings and infections, including symptoms such as headache, nausea, vomiting, dizziness, fatigue, and a feeling of weakness. Affected families often believe they are victims of food poisoning. Infants may be irritable and feed poorly. Neurological signs include confusion, disorientation, visual disturbance, syncope and seizures.

Some descriptions of carbon monoxide poisoning include retinal hemorrhages, and an abnormal cherry-red blood hue. In most clinical diagnoses these signs are seldom noticed. One difficulty with the usefulness of this cherry-red effect is that it corrects, or masks, what would otherwise be an unhealthy appearance, since the chief effect of removing deoxygenated hemoglobin is to make an asphyxiated person appear more normal, or a dead person appear more lifelike, similar to the effect of red colourants in embalming fluid. The "false" or unphysiologic red-colouring effect in anoxic CO-poisoned tissue is related to the meat-colouring commercial use of carbon monoxide, discussed below.

Carbon monoxide also binds to other molecules such as myoglobin and mitochondrial cytochrome oxidase. Exposures to carbon monoxide may cause significant damage to the heart and central nervous system, especially to the globus pallidus, often with long-term chronic pathological conditions. Carbon monoxide may have severe adverse effects on the fetus of a pregnant woman.

Normal Human Physiology

Carbon monoxide is produced naturally by the human body as a signalling molecule. Thus, carbon monoxide may have a physiological role in the body, such as a neurotransmitter or a blood vessel relaxant. Because of carbon monoxide's role in the body, abnormalities in its metabolism have been linked to a variety of diseases, including neurodegenerations, hypertension, heart failure, and inflammation.

Functional Summary

- CO functions as an endogenous signalling molecule.
- CO modulates functions of the cardiovascular system.
- CO inhibits blood platelet aggregation and adhesion.
- CO may play a role as potential therapeutic agent.

Microbiology

Carbon monoxide is a nutrient for methanogenic archaea, a building-block for acetylcoenzyme A. This is the theme for the emerging field of bioorganometallic chemistry. Extremophile micro-organisms can, thus, metabolise carbon monoxide in such locations as the thermal vents of volcanoes.

In bacteria, carbon monoxide is produced via the reduction of carbon dioxide by the enzyme carbon monoxide dehydrogenase, an Fe-Ni-S-containing protein.

CooA is a carbon monoxide sensor protein. The scope of its biological role is still unknown; it may be part of a signalling pathway in bacteria and archaea. Its occurrence in mammals is not established.

Occurrence

Carbon monoxide occurs in various natural and artificial environments. Typical concentrations in parts per million are as follows:

Composition of dry atmosphere, by volume

ppmv: parts per million by volume (note: volume fraction is equal to mole fraction for ideal gas only)

Concentration	*Source*
0.1 ppmv	Natural atmosphere level (MOPITT)0.5–5 ppmv Average level in homes5–15 ppmv Near-properly adjusted gas stoves in homes, modern vehicle exhaust emissions17 ppmv Atmosphere of Venus
100–200 ppmv	Exhaust from automobiles in the Mexico City central area700 ppmv Atmosphere of Mars
5,000 ppmv	Exhaust from a home wood fire7,000 ppmv Undiluted warm car exhaust without a catalytic converter

Atmospheric Presence

Carbon monoxide is present in small amounts in the atmosphere, chiefly as a product of volcanic activity but also from natural and man-made fires (such as forest and bushfires, burning of crop residues, and sugarcane fire-cleaning).

The burning of fossil fuels also contributes to carbon monoxide production. Carbon monoxide occurs dissolved in molten volcanic rock at high pressures in the Earth's mantle. Because natural sources of carbon monoxide are so variable from year to year, it is extremely difficult to accurately measure natural emissions of the gas.

Carbon monoxide has an indirect radiative forcing effect by elevating concentrations of methane and tropospheric ozone through chemical reactions with other atmospheric constituents (e.g., the hydroxyl radical, OH·) that would otherwise destroy them. Through natural processes in the atmosphere, it is eventually oxidized to carbon dioxide. Carbon monoxide concentrations are both short-lived in the atmosphere and spatially variable.

In the atmosphere of Venus carbon monoxide occurs as a result of the photodissociation of carbon dioxide by electromagnetic radiation of wavelengths shorter than 169 nm.

Due to its long lifetime in the mid-troposphere, carbon monoxide is also used as tracer of transport for pollutant plumes.

Urban Pollution

Carbon monoxide is a temporary atmospheric pollutant in some urban areas, chiefly from the exhaust of internal combustion engines (including vehicles, portable and back-up generators, lawn mowers, power washers, etc.), but also from incomplete combustion of various other fuels (including wood, coal, charcoal, oil, paraffin, propane, natural gas, and trash).

Large CO pollution events can be observed from space over cities.

Role in Ground-level Ozone Formation

Carbon monoxide is, along with aldehydes, part of the series of cycles of chemical reactions that form photochemical smog. It reacts with hydroxyl radical ($^{\cdot}OH$) to produce a radical intermediate $^{\cdot}HOCO$, which transfers rapidly its radical hydrogen to O_2 to form peroxy radical ($HO_2^{\cdot}$) and carbon dioxide (CO_2). Peroxy radical subsequently reacts with nitrogen oxide (NO) to form nitrogen dioxide (NO_2) and hydroxyl radical. NO_2 gives $O(^3P)$ via photolysis, thereby forming O_3 following reaction with O_2. Since hydroxyl radical is formed during the formation of NO_2, the balance of the sequence of chemical reactions starting with carbon monoxide and leading to the formation of ozone is:

$$CO + 2O_2 + h\nu \rightarrow CO_2 + O_3$$

(where $h\nu$ refers to the photon of light absorbed by the NO_2 molecule in the sequence)

Although the creation of NO_2 is the critical step leading to low level ozone formation, it also increases this ozone in another, somewhat mutually exclusive way, by reducing the quantity of NO that is available to react with ozone.

Indoor Pollution

In closed environments, the concentration of carbon monoxide can easily rise to lethal levels. On average, 170 people in the United States die every year from carbon monoxide produced by non-automotive consumer products. However, according to the Florida Department of Health, "every year more than 500 Americans die from accidental exposure to carbon monoxide and thousands more across the U.S. require emergency medical care for non-fatal carbon monoxide poisoning" These products include malfunctioning fuel-burning appliances such as furnaces, ranges, water heaters, and gas and kerosene room heaters; engine-powered equipment such as portable generators; fireplaces; and charcoal that is burned in homes and other enclosed areas. The American Association of Poison Control Centres (AAPCC) reported 15,769 cases of carbon monoxide poisoning resulting in 39 deaths in 2007. In 2005, the CPSC reported 94 generator-related carbon monoxide poisoning deaths.

Forty-seven of these deaths were known to have occurred during power outages due to severe weather, including Hurricane Katrina. Still others die from carbon monoxide produced by non-consumer products, such as cars left running in attached garages. The Centres for Disease Control and Prevention estimates that several thousand people go to hospital emergency rooms every year to be treated for carbon monoxide poisoning.

Blood Presence

Carbon monoxide is absorbed through breathing and enters the blood stream through gas exchange in the lungs. It is also produced in hemoglobim metabolism and enters the blood from the tissues, and thus is present in all normal tissues, even if not inhaled.

Normal circulating levels in the blood are 0% to 3%, and are higher in smokers. Carbon monoxide levels cannot be assessed through a physical exam. Laboratory testing requires a blood sample (arterial or venous) and laboratory analysis on a CO-Oximetre. Additionally, a noninvasive carboxyhemoglobin (SpCO) test method from Pulse CO-Oximetry exists and has been validated compared to invasive methods.

Astrophysics

Outside of Earth, carbon monoxide is the second-most common molecule in the interstellar medium, after molecular hydrogen. Because of its asymmetry, the carbon monoxide molecule produces far brighter spectral lines than the hydrogen molecule, making CO much easier to detect. Interstellar CO was first detected with radio telescopes in 1970. It is now the most commonly used tracer of molecular gas in general in the interstellar medium of galaxies, as molecular hydrogen can only be detected using ultraviolet light, which requires space telescopes. Carbon monoxide observations provide much of the information about the molecular clouds in which most stars form.

Beta Pictoris, the second brightest star in the constellation Pictor, shows an excess of infrared emission compared to normal stars of its type, which is caused by large quantities of dust and gas (including carbon monoxide) near the star.

Production

Many methods have been developed for carbon monoxide's production.

Industrial Production

A major industrial source of CO is producer gas, a mixture containing mostly carbon monoxide and nitrogen, formed by combustion of carbon in air at high temperature when there is an excess of carbon. In an oven, air is passed through a bed of coke. The initially produced CO_2 equilibrates with the remaining hot carbon to give CO. The reaction of CO_2 with carbon to give CO is described as the Boudouard reaction. Above 800 °C, CO is the predominant product:

$$CO_2 + C \rightarrow 2\,CO\ (\Delta H = 170\ kJ/mol)$$

Another source is "water gas", a mixture of hydrogen and carbon monoxide produced via the endothermic reaction of steam and carbon:

$$H_2O + C \rightarrow H_2 + CO\ (\Delta H = +131\ kJ/mol)$$

Other similar "synthesis gases" can be obtained from natural gas and other fuels.

Carbon monoxide is also a byproduct of the reduction of metal oxide ores with carbon, shown in a simplified form as follows:

$$MO + C \rightarrow M + CO$$

Since CO is a gas, the reduction process can be driven by heating, exploiting the positive (favourable) entropy of reaction. The Ellingham diagram shows that CO formation is favoured over CO_2 in high temperatures.

Laboratory Preparation

Carbon monoxide is conveniently produced in the laboratory by the dehydration of formic acid or oxalic acid, for example with concentrated sulphuric acid. Another method is heating an intimate mixture of powdered zinc metal and calcium carbonate, which releases CO and leaves behind zinc oxide and calcium oxide:

$$Zn + CaCO_3 \rightarrow ZnO + CaO + CO$$

Coordination Chemistry

Most metals form coordination complexes containing covalently attached carbon monoxide. Only metals in lower oxidation states will complex with carbon monoxide ligands. This is because there must be sufficient electron density to facilitate back-donation from the metal d_{xz}-orbital, to the π*molecular orbital from CO. The lone pair on the carbon atom in CO, also donates electron density to the $d_{x^2-y^2}$ on the metal to form a sigma bond. This electron donation is also exhibited with the cis effect, or the labilization of CO ligands in the cis position. Nickel carbonyl, for example, forms by the direct combination of carbon monoxide and nickel metal:

$$Ni + 4\ CO \rightarrow Ni(CO)_4\ (1\ bar,\ 55\ °C)$$

For this reason, nickel in any tubing or part must not come into prolonged contact with carbon monoxide. Nickel carbonyl decomposes readily back to Ni and CO upon contact with hot surfaces, and this method is used for the industrial purification of nickel in the Mond process.

In nickel carbonyl and other carbonyls, the electron pair on the carbon interacts with the metal; the carbon monoxide donates the electron pair to the metal. In these situations, carbon monoxide is called the carbonyl ligand. One of the most important metal carbonyls is iron pentacarbonyl, $Fe(CO)_5$:

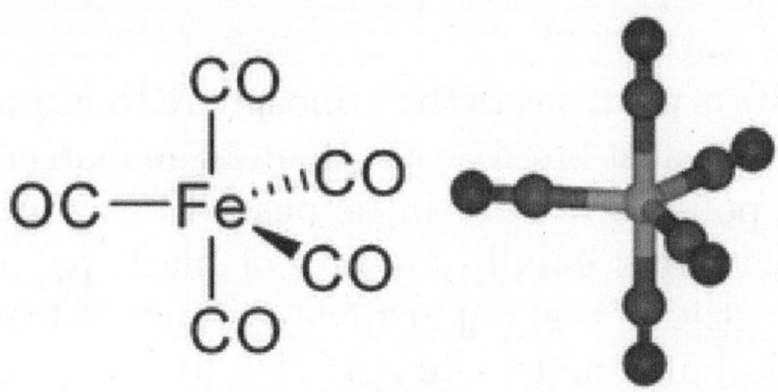

Many metal-CO complexes are prepared by decarbonylation of organic solvents, not from CO. For instance, iridium trichloride and triphenylphosphine react in boiling 2-methoxyethanol or DMF to afford $IrCl(CO)(PPh_3)_2$.

Metal carbonyls in coordination chemistry are usually studied using infrared spectroscopy.

Organic and Main Group Chemistry

In the presence of strong acids and water, carbon monoxide reacts with alkenes to form carboxylic acids in a process known as the Koch–Haaf reaction. In the Gattermann–Koch reaction, arenes are converted to benzaldehyde derivatives in the presence of $AlCl_3$ and HCl. Organolithium compounds (e.g. butyl lithium) react with carbon monoxide, but these reactions have little scientific use.

Although CO reacts with carbocations and carbanions, it is relatively nonreactive toward organic compounds without the intervention of metal catalysts.

With main group reagents, CO undergoes several noteworthy reactions. Chlorination of CO is the industrial route to the important compound phosgene. With borane CO forms an adduct, H_3BCO, which is isoelectronic with the acylium cation $[H_3CCO]^+$. CO reacts with sodium to give products resulting from C-C coupling such as sodium acetylenediolate $2Na^+ \cdot C_2O^{2-}{}_2$. It reacts with molten potassium to give a mixture of an organometallic compound, potassium acetylenediolate $2K^+ \cdot C_2O^{2-}{}_2$, potassium benzenehexolate $6K^+C_6O^{6-}{}_6$, and potassium rhodizonate $2K^+ \cdot C_6O^{2-}{}_6$.

The compounds cyclohexanehexone or triquinoyl (C_6O_6) and cyclopentanepentone or leuconic acid (C_5O_5), which so far have been obtained only in trace amounts, can be regarded as polymers of carbon monoxide.

At pressures of over 5 gigapascals, carbon monoxide converts into a solid polymer of carbon and oxygen. This is metastable at atmospheric pressure but is a powerful explosive.

Dust

Dust consists of particles in the atmosphere that come from various sources such as soil, dust lifted by weather (an aeolian process), volcanic eruptions, and pollution. Dust in homes, offices, and other human environments contains small amounts of plant pollen, human and animal hairs, textile fibres, paper fibres, minerals from outdoor soil, human skin cells, burnt meteorite particles, and many other materials which may be found in the local environment.

Domestic Dust and Humans

Dust may worsen hay fever. Circulating outdoor air through a house by keeping doors and windows open — or at least slightly ajar — may reduce the risk of hay-fever-causing dust. However, in colder climates, occupants seal even the smallest air gaps, which keeps fresh air from outside from circulating inside the house. Because of this, it is essential to have a plan for managing dust and airflow.

House dust mites are present indoors wherever humans live. Positive tests for dust mite allergies are extremely common among people with asthma. Dust mites are microscopic arachnids whose primary food is dead human skin cells, but they do not live on living people. They and their faeces and other allergens they produce are major constituents of house dust, but because they are so heavy they are not suspended for long in the air. They are generally found on the floor and other surfaces until disturbed (by walking, for example). It could take somewhere between twenty minutes and two hours for dust mites to settle back down out of the air.

Dust mites are a nesting species that prefers a dark, warm, and humid climate. They flourish in mattresses, bedding, upholstered furniture, and carpets. Their faeces include enzymes that are released upon contact with a moist surface, which can happen when a person inhales, and these enzymes can kill cells within the human body. House dust mites did not become a problem until humans began to use textiles, such as western style blankets and clothing.

Atmospheric Dust

Atmospheric or wind-borne dust, also known as *aeolian dust*, comes from arid and dry regions where high velocity winds are able to remove mostly silt-sized material, deflating susceptible surfaces. This includes areas where grazing, ploughing, vehicle use, and other human activities have further destabilized the land, though not all source areas have been largely affected by anthropogenic impacts. One-third of the global land area is covered by dust-producing surfaces, made up of hyper-arid regions like the Sahara which covers 0.9 billion hectares, and drylands which occupy 5.2 billion hectares.

Dust in the atmosphere is produced by saltation and sandblasting of sand-sized grains, and it is transported through the troposphere. This airborne dust is considered an aerosol and once in the atmosphere, it can produce strong local radiative forcing. Saharan dust in particular can be transported and deposited as far as the Caribbean and Amazonia, and may affect air temperatures, cause ocean cooling, and alter rainfall amounts. Coal dust is responsible for the lung disease known as pneumoconiosis, including black lung disease that occurs among coal miners. The danger of coal dust resulted in environmental legislation regulating work place air quality in some jurisdictions. In addition, if enough coal dust is dispersed within the air in a given area, it can create an explosion hazard under certain circumstances.

Middle East

Dust in the Middle East has been a historic phenomenon. Recently, because of climate change and the escalating process of desertification, the problem has worsened dramatically. As a multi-factor phenomenon, there is not yet a clear consensus on the sources or potential solutions to the problem.

In Iran, the dust is already affecting more than 5 million people directly, and has emerged as a serious government issue in recent years.

In the province of Khuzestan it has led to the severe reduction of air quality. The amount of pollutants in the air has surpassed more than 50 times the normal level several times in a year. Recently, initiatives such as Project-Dust have been established to directly study the Middle Eastern dust.

Road Dust

Dust kicked up by vehicles travelling on roads may make up 33% of air pollution. Road dust consists of deposition of vehicle exhausts and industrial exhausts, tire and brake wear, dust from paved roads or potholes, and dust from construction sites. Road dust is a significant source contributing to the generation and release of particulate matter into the atmosphere. Control of road dust is a significant challenge in urban areas, and also in other spheres with high levels of vehicular traffic upon unsealed roads such as mines and garbage dumps. Road dust may be suppressed by mechanical methods like sweeping vehicles, with vegetable oils, or with water sprayers. Improvements in automotive engineering have reduced the amount of PM10s produced by road traffic, the proportion representing re-suspension of existing particulates has, as a result, increased.

Dust control

Control of Atmospheric Dust

The U. S. Environmental Protection Agency (EPA) mandates facilities that generate dust minimize or mitigate the production of dust in their operation. The most frequent dust control violations occur at new residential housing developments in urban areas. United States Federal law requires that construction sites obtain permits to conduct earth moving, and include plans to control dust emissions. Control measures include such simple practices as spraying construction and demolition sites with water, and preventing the tracking of dust onto adjacent roads.

Some of the issues you may be facing may include;

- Reducing dust related health risks that include allergic reactions, pneumonia and asthmatic attacks.
- Improving visibility and road safety.
- Providing cleaner air, cleaner vehicles and cleaner homes and promoting better health.
- Improving crop productivity in agriculture.
- Reducing vehicle maintenance costs by lowering the levels of dust that clog filters, bearings and machinery.

- Reducing driver fatigue, maintenance on suspension systems and improving fuel economy.
- Increasing cumulative effect - each new application builds on previous residuals reducing re-application rate *while improving performance.

US federal laws require dust control on sources such as vacant lots, unpaved parking lots, and unpaved roads. Dust in such places may be suppressed by mechanical methods, including paving or laying down gravel, or stabilizing the surface with water, vegetable oils or other dust suppressants, or by using water misters to suppress dust that is already airborne.

Control of Domestic Dust

Dust control is the suppression of solid particles with diametres less than 500 micrometres. Dust in the airstream poses a serious health threat to children, older people, and those with respiratory illnesses.

House dust can become airborne easily. Care is required when removing dust to avoid causing the dust to become airborne. A feather duster tends to agitate the dust so it lands elsewhere. Products like Pledge and Swiffer are specifically made for removing dust by trapping it with sticky chemicals.

Certified HEPA (tested to MIL STD 282) can effectively trap 99.97% of dust at 0.3 micrometres. Not all HEPA (type/media) filters can effectively stop dust; while vacuum cleaners with HEPA (type/ media) filters, water, or cyclones may filter more effectively than without, they may still exhaust millions of particles per cubic foot of air circulated. Central vacuum cleaners can be effective in removing dust, especially if they are exhausted directly to the outdoors.

Air filtering appliances differ greatly in their effectiveness. Laser particle counters are an effective way to measure filter effectiveness, medical grade instruments can test for particles as small as 0.3 micrometres. In order to test for dust in the air, there are several options available. Pre weighted filter and matched weight filters made from polyvinyl chloride or mixed cellulose ester are suitable for respirable dust (less than 10 micrometres in diametre).

Control of Dust Resistance on Surfaces

A dust resistant surface is a state of prevention against dust contamination or damage, by a design or treatment of materials and items in manufacturing or through a repair process. A reduced tacticity of a synthetic layer or covering can protect surfaces and release small molecules that could have remained attached. A panel, container or enclosure with seams may feature types of strengthened rigidity or sealant to vulnerable edges and joins.

Water

Water is a transparent fluid which forms the world's streams, lakes, oceans and rain, and is the major constituent of the fluids of living things. As a chemical compound, a water molecule contains one oxygen and two hydrogen atoms that are connected by covalent bonds. Water is a liquid at standard ambient temperature and pressure, but it often co-exists on Earth with its solid state, ice; and gaseous state, steam (water vapour).

Water covers 71% of the Earth's surface. It is vital for all known forms of life. On Earth, 96.5% of the planet's water is found in seas and oceans, 1.7% in groundwater, 1.7% in glaciers and the ice caps of Antarctica and Greenland, a small fraction in other large water bodies, and 0.001% in the air as vapour, clouds (formed of solid and liquid water particles suspended in air), and precipitation. Only 2.5% of the Earth's water is freshwater, and 98.8% of that water is in ice and groundwater. Less than 0.3% of all freshwater is in rivers, lakes, and the atmosphere, and an even smaller amount of the Earth's freshwater (0.003%) is contained within biological bodies and manufactured products.

Water on Earth moves continually through the water cycle of evaporation and transpiration (evapotranspiration), condensation, precipitation, and runoff, usually reaching the sea. Evaporation and transpiration contribute to the precipitation over land. Water used in the production of a good or service is known as virtual water.

Safe drinking water is essential to humans and other lifeforms even though it provides no calories or organic nutrients. Access to safe drinking water has improved over the last decades in almost every part of the world, but approximately one billion people still lack access to safe water and over 2.5 billion lack access to adequate sanitation. There is a clear correlation between access to safe water and gross domestic product per capita. However, some observers have estimated that by 2025 more than half of the world population will be facing water-based vulnerability. A report, issued in November 2009, suggests that by 2030, in some developing regions of the world, water demand will exceed supply by 50%. Water plays an important role in the world economy, as it functions as a solvent for a wide variety of chemical substances and facilitates industrial cooling and transportation.

Approximately 70% of the fresh water used by humans goes to agriculture.

Chemical and Physical Properties

Water is the chemical substance with chemical formula H2O: one molecule of water has two hydrogen atoms covalently bonded to a single oxygen atom.

Water appears in nature in all three common states of matter (solid, liquid, and gas) and may take many different forms on Earth: water vapour and clouds in the sky, seawater in the oceans, icebergs in the polar oceans, glaciers in the mountains, fresh and salt water lakes, rivers, and aquifers in the ground.

The major chemical and physical properties of water are:

- Water is a liquid at standard temperature and pressure. It is tasteless and odourless. The intrinsic colour of water and ice is a very slight blue hue, although both appear colourless in small quantities. Water vapour is essentially invisible as a gas.
- Water is transparent in the visible electromagnetic spectrum. Thus aquatic plants can live in water because sunlight can reach them. Infrared light is strongly absorbed by the hydrogen-oxygen or OH bonds.
- Since the water molecule is not linear and the oxygen atom has a higher electronegativity than hydrogen atoms, the oxygen atom carries a slight negative charge, whereas the hydrogen atoms are slightly positive. As a result, water is a polar molecule with an electrical dipole moment. Water also can form an unusually large number of intermolecular hydrogen bonds (four) for a molecule of its size. These factors lead to strong attractive forces between molecules of water, giving rise to water's high surface tension and capillary forces. The capillary action refers to the tendency of water to move up a narrow tube against the force of gravity. This property is relied upon by all vascular plants, such as trees.
- Water is a good polar solvent and is often referred to as *the universal solvent*. Substances that dissolve in water, e.g., salts, sugars, acids, alkalis, and some gases – especially oxygen and carbon dioxide (carbonation) – are known as *hydrophilic* (water-loving) substances, while those that are immiscible with water (e.g., fats and oils), are known as *hydrophobic* (water-fearing) substances.

- All of the components in cells (proteins, DNA and polysaccharides) are dissolved in water, deriving their structure and activity from their interactions with the water.
- Pure water has a low electrical conductivity, but this increases with the dissolution of a small amount of ionic material such as sodium chloride.
- The boiling point of water (and all other liquids) is dependent on the barometric pressure. For example, on the top of Mount Everest water boils at 68 °C (154 °F), compared to 100 °C (212 °F) at sea level at a similar latitude (since latitude modifies atmospheric pressure slightly). Conversely, water deep in the ocean near geothermal vents can reach temperatures of hundreds of degrees and remain liquid.
- At 4181.3 J/(kg·K), water has a high specific heat capacity, as well as a high heat of vaporization (40.65 kJ·mol^{-1}), both of which are a result of the extensive hydrogen bonding between its molecules. These two unusual properties allow water to moderate Earth's climate by buffering large fluctuations in temperature.
- The density of liquid water is 1,000 kg/m^3 (62.43 lb/cu ft) at 4 °C. Ice has a density of 917 kg/m^3 (57.25 lb/cu ft).
- The maximum density of water occurs at 3.98 °C (39.16 °F). Most known pure substances become more dense as they cool, however water has the anomalous property of becoming less dense when it is cooled to its solid form, ice. During cooling water becomes more dense until reaching 3.98 °C. Below this temperature, the open structure of ice is gradually formed in the low temperature water; the random orientations of the water molecules in the liquid are maintained by the thermal motion, and below 3.98 °C there is not enough thermal energy to maintain this randomness. As water is cooled there are two competing effects: 1) decreasing volume, and 2) increase overall volume of the liquid as the molecules begin to orient into the organized structure of ice. Between 3.98 °C and 0 °C, the second effect will cancel the first effect so the net effect is an increase of volume with decreasing temperature. Water expands to occupy a 9% greater volume as ice, which accounts for the fact that ice floats on liquid water, as in icebergs.
- Water is miscible with many liquids, such as ethanol, in all proportions, forming a single homogeneous liquid. On the other

hand, water and most oils are immiscible, usually forming layers with the least dense liquid as the top layer, and the most dense layer at the bottom.

- Water forms an azeotrope with many other solvents.
- Water can be split by electrolysis into hydrogen and oxygen. The energy required to split water into hydrogen and oxygen by electrolysis or any other means is greater than the energy that can be collected when the hydrogen and oxygen recombine.
- As an oxide of hydrogen, water is formed when hydrogen or hydrogen-containing compounds burn or react with oxygen or oxygen-containing compounds. Water is not a fuel, it is an end-product of the combustion of hydrogen.
- Elements which are more electropositive than hydrogen such as lithium, sodium, calcium, potassium and caesium displace hydrogen from water, forming hydroxides. Being a flammable gas, the hydrogen given off is dangerous and the reaction of water with the more electropositive of these elements may be violently explosive.

Property	*Remarks*	*Importance to the environment*
Physical state	Only substance occurring naturally in all three phases as solid, liquid, and gas on Earth's surface	Transfer of heat between ocean and atmosphere by phase change
Dissolving ability	Dissolves more substances in greater quantities than any other common liquid	Important in chemical, physical, and biological processes
Density: mass per unit volume	Density is determined by (1) temperature, (2) salinity, and (3) pressure, in that order of importance. The temperature of maximum density for pure water is 4 °C. For seawater, the freezing point decreases with increasing salinity	Controls oceanic vertical circulation, aids in heat distribution, and allows seasonal stratification
Surface tension	Highest of all common liquids	Controls drop formation in rain and clouds; important in cell physiology
Conduction of heat	Highest of all common liquids	Important on the small scale, especially on cellular level
Heat capacity	Highest of all common solids and liquids	Prevents extreme range in Earth's temperatures (i.e., great heat moderator)
Latent heat of fusion	Highest of all common liquids and most solids	Thermostatic heat-regulating effect due to the release of heat on freezing and absorption on melting
Latent heat of vaporization	Highest of all common substances	Immense importance: a major factor in the transfer of heat in and between ocean and atmosphere, driving weather and climate
Refractive index	Increases with increasing salinity and decreases with increasing temperature	Objects appear closer than in air
Transparency	Relatively great for visible light; absorption high for infrared and ultraviolet	Important for photosynthesis
Sound transmission	Good compared with other fluids	Allows for sonar and precision depth recorders to rapidly determine water depth, and to detect subsurface features and animals; sounds can be heard great distances underwater
Compressibility	Only slight	Density changes only slightly with pressure/depth
Boiling and melting points	Unusually high	Allows water to exist as a liquid on most of Earth

Taste and Odour

Pure H_2O is tasteless and odourless. Water can dissolve many different substances, giving it varying tastes and odors. Humans, and other animals, have developed senses that enable them to evaluate the potability of water by avoiding water that is too salty or putrid.

The taste of spring water and mineral water, often advertised in marketing of consumer products, derives from the minerals dissolved in it. The advertised purity of spring and mineral water refers to absence of toxins, pollutants, and microbes, not to the absence of naturally occurring minerals.

Distribution in Nature

In the Universe

Much of the universe's water is produced as a byproduct of star formation. When stars are born, their birth is accompanied by a strong outward wind of gas and dust. When this outflow of material eventually impacts the surrounding gas, the shock waves that are created compress and heat the gas. The water observed is quickly produced in this warm dense gas.

On 22 July 2011 a report described the discovery of a gigantic cloud of water vapour containing –140 trillion times more water than all of Earth's oceans combined– around a quasar located 12 billion light years from Earth. According to the researchers, the "discovery shows that water has been prevalent in the universe for nearly its entire existence".

Water has been detected in interstellar clouds within our galaxy, the Milky Way. Water probably exists in abundance in other galaxies, too, because its components, hydrogen and oxygen, are among the most abundant elements in the universe. Based on models of the formation and evolution of the Solar System and that of other star systems, most other planetary systems are likely to have similar ingredients.

Water Vapour

Water is present as vapour in:

- Atmosphere of the Sun: in detectable trace amounts
- Atmosphere of Mercury: 3.4%, and large amounts of water in Mercury's exosphere
- Atmosphere of Venus: 0.002%
- Earth's atmosphere: ~0.40% over full atmosphere, typically 1–4% at surface; as well as that of the Moon in trace amounts

- Atmosphere of Mars: 0.03%
- Atmosphere of Ceres
- Atmosphere of Jupiter: 0.0004% – in ices only; and that of its moon Europa
- Atmosphere of Saturn – in ices only; and that of its moons Titan (stratospheric), Enceladus: 91% and Dione (exosphere)
- Atmosphere of Uranus - in trace amounts below 50 bar
- Atmosphere of Neptune - found in the deeper layers
- Extrasolar planet atmospheres: including those of HD 189733 b and HD 209458 b, Tau Boötis b and HAT-P-11b
- Circumstellar disks: including those of TW Hydrae, IRC +10216 and APM 08279+5255

Liquid Water

Liquid water is known to be present on Earth, covering 71% of our planet's surface. Scientists believe liquid water is present in the Saturnian moons of Enceladus, as a 10 kilometre thick ocean approximately 30-40 kilometres below Enceladus' south polar surface, and Titan, as a subsurface layer, possibly mixed with ammonia. Liquid water may also exist on Jupiter's moon Ganymede as a layer sandwiched between high pressure ice and rock.

Water Ice

Water is present as ice on:

- Mars: under the regolith and at the poles
- Earth-Moon system: mainly as ice sheets on Earth and in Lunar craters and volcanic rocks NASA reported the detection of water molecules by NASA's Moon Mineralogy Mapper aboard the Indian Space Research Organization's Chandrayaan-1 spacecraft in September 2009.
- Jupiter's moons: Europa's surface and also that of Ganymede
- Saturn: in the planets ring system and on the surface and mantle of Titan and Enceladus
- Pluto-Charon system
- Comets and related (Kuiper belt and Oort cloud objects).

And may also be present on:

- Mercury's poles
- Ceres
- Tethys

Exotic Forms

Water and other volatiles probably comprise much of the internal structures of Uranus and Neptune and the water in the deeper layers may be in the form of ionic water in which the molecules break down into a soup of hydrogen and oxygen ions, and deeper down as superionic water in which the oxygen crystallises but the hydrogen ions float around freely within the oxygen lattice.

Water and Habitable Zone

The existence of liquid water, and to a lesser extent its gaseous and solid forms, on Earth are vital to the existence of life on Earth as we know it. The Earth is located in the habitable zone of the solar system; if it were slightly closer to or farther from the Sun (about 5%, or about 8 million kilometres), the conditions which allow the three forms to be present simultaneously would be far less likely to exist.

Earth's gravity allows it to hold an atmosphere. Water vapour and carbon dioxide in the atmosphere provide a temperature buffer (greenhouse effect) which helps maintain a relatively steady surface temperature. If Earth were smaller, a thinner atmosphere would allow temperature extremes, thus preventing the accumulation of water except in polar ice caps (as on Mars). The surface temperature of Earth has been relatively constant through geologic time despite varying levels of incoming solar radiation (insolation), indicating that a dynamic process governs Earth's temperature via a combination of greenhouse gases and surface or atmospheric albedo. This proposal is known as the *Gaia hypothesis.*

The state of water on a planet depends on ambient pressure, which is determined by the planet's gravity. If a planet is sufficiently massive, the water on it may be solid even at high temperatures, because of the high pressure caused by gravity, as it was observed on exoplanets Gliese 436 b and GJ 1214 b.

On Earth

Hydrology is the study of the movement, distribution, and quality of water throughout the Earth. The study of the distribution of water is hydrography. The study of the distribution and movement of groundwater is hydrogeology, of glaciers is glaciology, of inland waters is limnology and distribution of oceans is oceanography. Ecological processes with hydrology are in focus of ecohydrology. The collective mass of water found on, under, and over the surface of a planet is called the hydrosphere. Earth's approximate water volume (the total water supply of the world) is 1,338,000,000 km^3 (321,000,000 mi^3).

Liquid water is found in bodies of water, such as an ocean, sea, lake, river, stream, canal, pond, or puddle. The majority of water on Earth is sea water. Water is also present in the atmosphere in solid, liquid, and vapour states. It also exists as groundwater in aquifers.

Water is important in many geological processes. Groundwater is present in most rocks, and the pressure of this groundwater affects patterns of faulting. Water in the mantle is responsible for the melt that produces volcanoes at subduction zones. On the surface of the Earth, water is important in both chemical and physical weathering processes. Water, and to a lesser but still significant extent, ice, are also responsible for a large amount of sediment transport that occurs on the surface of the earth. Deposition of transported sediment forms many types of sedimentary rocks, which make up the geologic record of Earth history.

Water Cycle

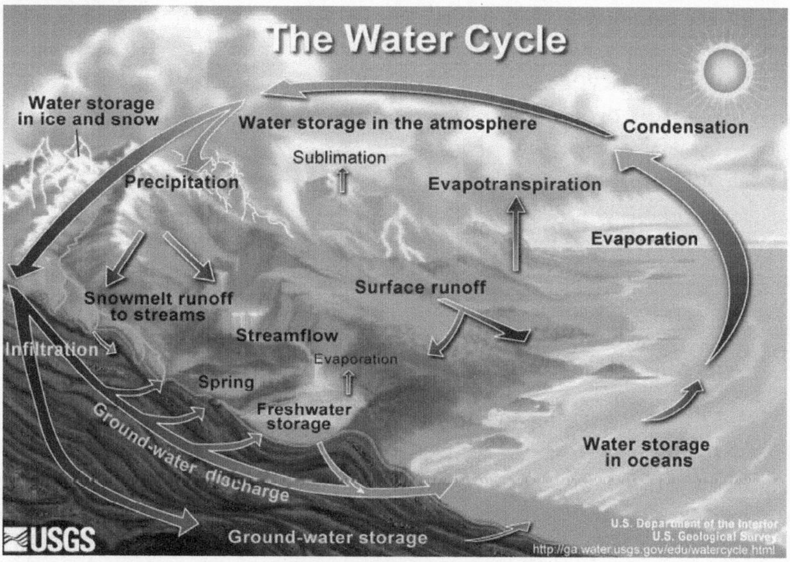

Figure: *Water cycle*

The water cycle (known scientifically as the hydrologic cycle) refers to the continuous exchange of water within the hydrosphere, between the atmosphere, soil water, surface water, groundwater, and plants.

Water moves perpetually through each of these regions in the *water cycle* consisting of following transfer processes:

- evaporation from oceans and other water bodies into the air and transpiration from land plants and animals into air.

- precipitation, from water vapour condensing from the air and falling to earth or ocean.
- runoff from the land usually reaching the sea.

Most water vapour over the oceans returns to the oceans, but winds carry water vapour over land at the same rate as runoff into the sea, about 47 Tt per year. Over land, evaporation and transpiration contribute another 72 Tt per year. Precipitation, at a rate of 119 Tt per year over land, has several forms: most commonly rain, snow, and hail, with some contribution from fog and dew. Dew is small drops of water that are condensed when a high density of water vapour meets a cool surface. Dew usually forms in the morning when the temperature is the lowest, just before sunrise and when the temperature of the earth's surface starts to increase. Condensed water in the air may also refract sunlight to produce rainbows.

Water runoff often collects over watersheds flowing into rivers. A mathematical model used to simulate river or stream flow and calculate water quality parametres is a hydrological transport model. Some water is diverted to irrigation for agriculture. Rivers and seas offer opportunity for travel and commerce. Through erosion, runoff shapes the environment creating river valleys and deltas which provide rich soil and level ground for the establishment of population centres. A flood occurs when an area of land, usually low-lying, is covered with water. It is when a river overflows its banks or flood comes from the sea. A drought is an extended period of months or years when a region notes a deficiency in its water supply. This occurs when a region receives consistently below average precipitation.

Fresh Water Storage

Some runoff water is trapped for periods of time, for example in lakes. At high altitude, during winter, and in the far north and south, snow collects in ice caps, snow pack and glaciers. Water also infiltrates the ground and goes into aquifers. This groundwater later flows back to the surface in springs, or more spectacularly in hot springs and geysers. Groundwater is also extracted artificially in wells. This water storage is important, since clean, fresh water is essential to human and other land-based life. In many parts of the world, it is in short supply.

Sea Water

Sea water contains about 3.5% salt on average, plus smaller amounts of other substances. The physical properties of sea water differ from fresh water in some important respects. It freezes at a lower temperature (about –1.9 °C) and its density increases with decreasing temperature to the freezing point, instead of reaching maximum density at a temperature above freezing. The salinity of water in major seas varies from about 0.7% in the Baltic Sea to 4.0% in the Red Sea.

Tides

Tides are the cyclic rising and falling of local sea levels caused by the tidal forces of the Moon and the Sun acting on the oceans. Tides cause changes in the depth of the marine and estuarine water bodies and produce oscillating currents known as tidal streams. The changing tide produced at a given location is the result of the changing positions of the Moon and Sun relative to the Earth coupled with the effects of Earth rotation and the local bathymetry. The strip of seashore that is submerged at high tide and exposed at low tide, the intertidal zone, is an important ecological product of ocean tides.

Agriculture

The most important use of water in agriculture is for irrigation, which is a key component to produce enough food. Irrigation takes up to 90% of water withdrawn in some developing countries and significant proportions in more economically developed countries (United States, 30% of freshwater usage is for irrigation).

Fifty years ago, the common perception was that water was an infinite resource. At this time, there were fewer than half the current number of people on the planet. People were not as wealthy as today, consumed fewer calories and ate less meat, so less water was needed to produce their food. They required a third of the volume of water we presently take from rivers. Today, the competition for the fixed amount of water resources is much more intense, giving rise to the concept of peak water. This is because there are now nearly seven billion people on the planet, their consumption of water-thirsty meat and vegetables is rising, and there is increasing competition for water from industry, urbanisation and biofuel crops. In future, even more water will be needed to produce food because the Earth's population is forecast to rise to 9 billion by 2050.

An assessment of water management in agriculture was conducted in 2007 by the International Water Management Institute in Sri Lanka to see if the world had sufficient water to provide food for its growing population. It assessed the current availability of water for agriculture on a global scale and mapped out locations suffering from water scarcity. It found that a fifth of the world's people, more than 1.2 billion, live in areas of physical water scarcity, where there is not enough water to meet all demands. A further 1.6 billion people live in areas experiencing economic water scarcity, where the lack of investment in water or insufficient human capacity make it impossible for authorities to satisfy the demand for water. The report found that it would be possible to produce the food required in future, but that continuation of today's food production and environmental trends would lead to crises in many parts of the world. To avoid a global water crisis, farmers will have to strive to increase productivity to meet growing demands for food, while industry and cities find ways to use water more efficiently.

Ice

Ice, solid substance produced by the freezing of water vapour or liquid water. At temperatures below 0 °C (32 °F), water vapour develops into frost at ground level and snowflakes (each of which consists of a single ice crystal) in clouds. Below the same temperature, liquid water forms a solid, as, for example, river ice, sea ice, hail, and ice produced commercially or in household refrigerators.

Ice occurs on the Earth's continents and surface waters in a variety of forms. Most notable are the continental glaciers (ice sheets) that cover much of Antarctica and Greenland. Smaller masses of perennial ice called ice caps occupy parts of Arctic Canada and other high-latitude regions, and mountain glaciers occur in more restricted areas, such as mountain valleys and the flatlands below.

Other occurrences of ice on land include the different types of ground ice associated with permafrost—that is, permanently frozen soil common to very cold regions. In the oceanic waters of the polar regions, icebergs occur when large masses of ice break off from glaciers or ice shelves and drift away. The freezing of seawater in these regions results in the formation of sheets of sea ice known as pack ice. During the winter months similar ice bodies form on lakes and rivers in many parts of the world. This article treats the structure and properties of ice in general. Ice in lakes and rivers, glaciers, icebergs, pack ice, and permafrost are treated separately in articles under their respective titles. For a detailed account of the widespread occurrences of glacial ice during the Earth's past.

Structure

The Water Molecule

Ice is the solid state of water, a normally liquid substance that freezes to the solid state at temperatures of 0 °C (32 °F) or lower and expands to the gaseous state at temperatures of 100 °C (212 °F) or higher. Water is an extraordinary substance, anomalous in nearly all its physical and chemical properties and easily the most complex of all the familiar substances that are single-chemical compounds. Consisting of two atoms of hydrogen (H) and one atom of oxygen (O), the water molecule has the chemical formula H_2O. These three atoms are covalently bonded (i.e., their nuclei are linked by attraction to shared electrons) and form a specific structure, with the oxygen atom located between the two hydrogen atoms. The three atoms do not lie in a straight line, however. Instead, the hydrogen atoms are bent toward each other, forming an angle of about 105°.

The three-dimensional structure of the water molecule can be pictured as a tetrahedron with an oxygen nucleus centre and four legs of high electron probability. The two legs in which the hydrogen nuclei are present are called bonding orbitals. Opposite the bonding orbitals and directed to the opposite corners of the tetrahedron are two legs of negative electrical charge. Known as the lone-pair orbitals, these are the keys to water's peculiar behaviour, in that they attract the hydrogen nuclei of adjacent water molecules to form what are called hydrogen bonds. These bonds are not especially strong, but, because they orient the water molecules into a specific configuration, they significantly affect the properties of water in its solid, liquid, and gaseous states.

In the liquid state, most water molecules are associated in a polymeric structure—that is, chains of molecules connected by weak hydrogen bonds. Under the influence of thermal agitation, there is a constant breaking and reforming of these bonds. In the gaseous state, whether steam or water vapour, water molecules are largely independent of one another, and, apart from collisions, interactions between them are slight. Gaseous water, then, is largely monomeric—i.e., consisting of single molecules—although there occasionally occur dimers (a union of two molecules) and even some trimers (a combination of three molecules).

In the solid state, at the other extreme, water molecules interact with one another strongly enough to form an ordered crystalline structure, with each oxygen atom collecting the four nearest of its neighbours and arranging them about itself in a rigid lattice. This structure results in a more open assembly, and hence a lower density, than the closely packed assembly of molecules in the liquid phase. For this reason, water is one of the few substances that is actually less dense in solid form than in the liquid state, dropping from 1,000 to 917 kilograms per cubic metre. It is the reason why ice floats rather than sinking, so that, during the winter, it develops as a sheet on the surface of lakes and rivers rather than sinking below the surface and accumulating from the bottom.

As water is warmed from the freezing point of 0 to 4 °C (from 32 to 39 °F), it contracts and becomes denser. This initial increase in density takes place because at 0°C a portion of the water consists of open-structured molecular arrangements similar to those of ice crystals. As the temperature increases, these structures break down and reduce their volume to that of the more closely packed polymeric structures of the liquid state. With further warming beyond 4°C, the water begins to expand in volume, along with the usual increase in intermolecular vibrations caused by thermal energy.

The Ice Crystal

At standard atmospheric pressure and at temperatures near 0 °C, the ice crystal commonly takes the form of sheets or planes of oxygen atoms joined in a series of open hexagonal rings. The axis parallel to the hexagonal rings is termed the *c*-axis and coincides with the optical axis of the crystal structure.

When viewed perpendicular to the *c*-axis, the planes appear slightly dimpled. The planes are stacked in a laminar structure that occasionally deforms by gliding, like a deck of cards. When this gliding deformation occurs, the bonds between the layers break, and the hydrogen atoms involved in those bonds must become attached to different oxygen atoms. In doing so, they migrate within the lattice, more rapidly at higher temperatures. Sometimes they do not reach the usual arrangement of two hydrogen atoms connected by covalent bonds to each oxygen atom, so that some oxygen atoms have only one or as many as three hydrogen bonds. Such oxygen atoms become the sites of electrical charge. The speed of crystal deformation depends on these readjustments, which in turn are sensitive to temperature. Thus the mechanical, thermal, and electrical properties of ice are interrelated.

Properties

Mechanical Properties

Like any other crystalline solid, ice subject to stress undergoes elastic deformation, returning to its original shape when the stress ceases. However, if a shear stress or force is applied to a sample of ice for a long time, the sample will first deform elastically and will then continue to deform plastically, with a permanent alteration of shape. This plastic deformation, or creep, is of great importance to the study of glacier flow. It involves two processes: intracrystalline gliding, in which the layers within an ice crystal shear parallel to each other without destroying the continuity of the crystal lattice, and recrystallization, in which crystal boundaries change in size or shape depending on the orientation of the adjacent crystals and the stresses exerted on them. The motion of dislocations—that is, of defects or disorders in the crystal lattice—controls the speed of plastic deformation. Dislocations do not move under elastic deformation.

The strength of ice, which depends on many factors, is difficult to measure. If ice is stressed for a long time, it deforms by plastic flow and has no yield point (at which permanent deformation begins) or ultimate strength. For short-term experiments with conventional testing machines, typical strength values in bars are 38 for crushing, 14 for bending, 9 for tensile, and 7 for shear.

Thermal Properties

The heat of fusion (heat absorbed on melting of a solid) of water is 334 kilojoules per kilogram. The specific heat of ice at the freezing point is 2.04 kilojoules per kilogram per degree Celsius. The thermal conductivity at this temperature is 2.24 watts per metre kelvin.

Another property of importance to the study of glaciers is the lowering of the melting point due to hydrostatic pressure: 0.0074 °C per bar. Thus for a glacier 300 metres (984 feet) thick, everywhere at the melting temperature, the ice at the base is 0.25 °C (0.45 °F) colder than at the surface.

Optical Properties

Pure ice is transparent, but air bubbles render it somewhat opaque. The absorption coefficient, or rate at which incident radiation decreases with depth, is about 0.1 cm^{-1} for snow and only 0.001 cm^{-1} or less for clear ice. Ice is weakly birefringent, or doubly refracting, which means that light is transmitted at different speeds in different crystallographic directions. Thin sections of snow or ice therefore can be conveniently studied under polarized light in much the same way that rocks are studied. The ice crystal strongly absorbs light in the red wavelengths, and thus the scattered light seen emerging from glacier crevasses and unweathered ice faces appears as blue or green.

Electromagnetic Properties

The albedo, or reflectivity (an albedo of 0 means that there is no reflectivity), to solar radiation ranges from 0.5 to 0.9 for snow, 0.3 to 0.65 for firn, and 0.15 to 0.35 for glacier ice. At the thermal infrared wavelengths, snow and ice are almost perfectly "black" (absorbent), and the albedo is less than 0.01. This means that snow and ice can either absorb or radiate long-wavelength radiation with high efficiency. At longer electromagnetic wavelengths (microwave and radio frequencies), dry snow and ice are relatively transparent, although the presence of even small amounts of liquid water greatly modifies this property. Radio echo sounding (radar) techniques are now used routinely to measure the thickness of dry polar glaciers, even where they are kilometres in thickness, but the slightest amount of liquid water distributed through the mass creates great difficulties with the technique.

Compositional layering of the Atmosphere

The atmosphere can be divided into two spheres on the basis of its chemical composition occurring with height i.e. (1) Homosphere. (2) Hereto sphere.

Homosphere: In the lower region up to the height of 88 km the various gases are thoroughly mixed and are homogenous by the process of turbulent mixing, and diffusion. This sphere is called as Homosphere. Herein the presence of gases is governed by the diffusion and the composition remains normally.

Chapter 6

Atmospheric Chemistry

Atmospheric chemistry is a branch of atmospheric science in which the chemistry of the Earth's atmosphere and that of other planets is studied.

It is a multidisciplinary field of research and draws on environmental chemistry, physics, meteorology, computer modelling, oceanography, geology and volcanology and other disciplines.

Research is increasingly connected with other areas of study such as climatology.

The composition and chemistry of the atmosphere is of importance for several reasons, but primarily because of the interactions between the atmosphere and living organisms.

The composition of the Earth's atmosphere changes as result of natural processes such as volcano emissions, lightning and bombardment by solar particles from corona.

It has also been changed by human activity and some of these changes are harmful to human health, crops and ecosystems.

Examples of problems which have been addressed by atmospheric chemistry include acid rain, ozone depletion, photochemical smog, greenhouse gases and global warming.

Atmospheric chemists seek to understand the causes of these problems, and by obtaining a theoretical understanding of them, allow possible solutions to be tested and the effects of changes in government policy evaluated.

Atmospheric Composition

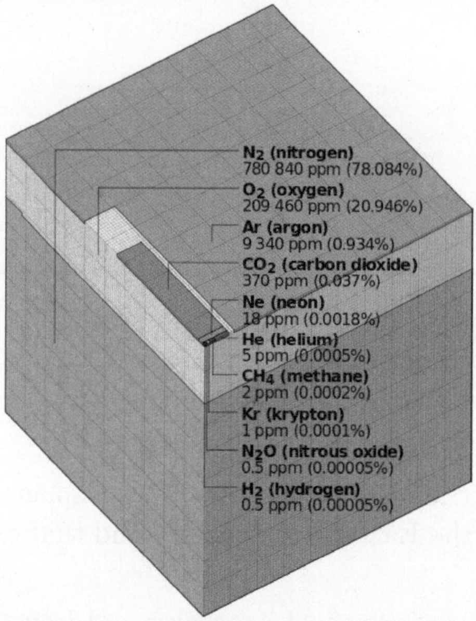

Figure: *Visualisation of composition by volume of Earth's atmosphere. Water vapour is not included as it is highly variable. Each tiny cube (such as the one representing krypton) has one millionth of the volume of the entire block. Data is from NASA Langley.*

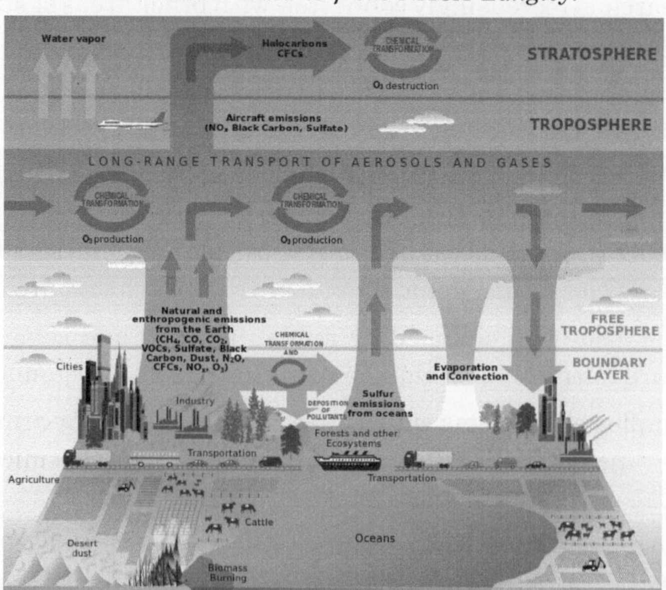

Figure: *Schematic of chemical and transport processes related to atmospheric composition.*

Average composition of dry atmosphere (mole fractions)	
Gas	per NASA
Nitrogen, N_2	78.084%
Oxygen, O_2	20.946%
Argon, Ar	0.934%
Minor constituents (mole fractions in ppm)	
Carbon Dioxide, CO_2	400
Neon, Ne	18.18
Helium, He	5.24
Methane, CH_4	1.7
Krypton, Kr	1.14
Hydrogen, H_2	0.55
	Water
Water vapour	Highly variable; typically makes up about 1%

Notes: the concentration of CO_2 and CH_4 vary by season and location. The mean molecular mass of air is 28.97 g/mol.

History

The ancient Greeks regarded air as one of the four elements, but the first scientific studies of atmospheric composition began in the 18th century. Chemists such as Joseph Priestley, Antoine Lavoisier and Henry Cavendish made the first measurements of the composition of the atmosphere.

In the late 19th and early 20th centuries interest shifted towards trace constituents with very small concentrations. One particularly important discovery for atmospheric chemistry was the discovery of ozone by Christian Friedrich Schönbein in 1840.

In the 20th century atmospheric science moved on from studying the composition of air to a consideration of how the concentrations of trace gases in the atmosphere have changed over time and the chemical processes which create and destroy compounds in the air.

Two particularly important examples of this were the explanation by Sydney Chapman and Gordon Dobson of how the ozone layer is created and maintained, and the explanation of photochemical smog by Arie Jan Haagen-Smit.

Further studies on ozone issues led to the 1995 Nobel Prize in Chemistry award shared between Paul Crutzen, Mario Molina and Frank Sherwood Rowland.

In the 21st century the focus is now shifting again. Atmospheric chemistry is increasingly studied as one part of the Earth system. Instead of concentrating on atmospheric chemistry in isolation the focus is now on seeing it as one part of a single system with the rest of the atmosphere, biosphere and geosphere. An especially important driver for this is the links between chemistry and climate such as the effects of changing climate on the recovery of the ozone hole and vice versa but also interaction of the composition of the atmosphere with the oceans and terrestrial ecosystems.

Methodology

Observations, lab measurements and modelling are the three central elements in atmospheric chemistry. Progress in atmospheric chemistry is often driven by the interactions between these components and they form an integrated whole. For example observations may tell us that more of a chemical compound exists than previously thought possible. This will stimulate new modelling and laboratory studies which will increase our scientific understanding to a point where the observations can be explained.

Observation

Observations of atmospheric chemistry are essential to our understanding. Routine observations of chemical composition tell us about changes in atmospheric composition over time. One important example of this is the Keeling Curve - a series of measurements from 1958 to today which show a steady rise in of the concentration of carbon dioxide. Observations of atmospheric chemistry are made in observatories such as that on Mauna Loa and on mobile platforms such as aircraft (e.g. the UK's Facility for Airborne Atmospheric Measurements), ships and balloons. Observations of atmospheric composition are increasingly made by satellites with important instruments such as GOME and MOPITT giving a global picture of air pollution and chemistry. Surface observations have the advantage that they provide long term records at high time resolution but are limited in the vertical and horizontal space they provide observations from. Some surface based instruments e.g. LIDAR can provide concentration profiles of chemical compounds and aerosol but are still restricted in the horizontal region they can cover. Many observations are available on line in Atmospheric Chemistry Observational Databases.

Lab Measurements

Measurements made in the laboratory are essential to our understanding of the sources and sinks of pollutants and naturally occurring compounds. Lab studies tell us which gases react with each other and how fast they react. Measurements of interest include reactions in the gas phase, on surfaces and in water. Also of high importance is photochemistry which quantifies how quickly molecules are split apart by sunlight and what the products are plus thermodynamic data such as Henry's law coefficients.

Modelling

In order to synthesise and test theoretical understanding of atmospheric chemistry, computer models (such as chemical transport models) are used. Numerical models solve the differential equations governing the concentrations of chemicals in the atmosphere. They can be very simple or very complicated. One common trade off in numerical models is between the number of chemical compounds and chemical reactions modelled versus the representation of transport and mixing in the atmosphere. For example, a box model might include hundreds or even thousands of chemical reactions but will only have a very crude representation of mixing in the atmosphere. In contrast, 3D models represent many of the physical processes of the atmosphere but due to constraints on computer resources will have far fewer chemical reactions and compounds.

Models can be used to interpret observations, test understanding of chemical reactions and predict future concentrations of chemical compounds in the atmosphere. One important current trend is for atmospheric chemistry modules to become one part of earth system models in which the links between climate, atmospheric composition and the biosphere can be studied. Some models are constructed by automatic code generators (e.g. Autochem or KPP). In this approach a set of constituents are chosen and the automatic code generator will then select the reactions involving those constituents from a set of reaction databases. Once the reactions have been chosen the ordinary differential equations (ODE) that describe their time evolution can be automatically constructed.

Hydrosphere

Igor Shiklomanov, the man selected by the United Nations to do its world inventory of water resources, estimated that there is 1386 million cubic kilometres of water on earth. This includes water in liquid and frozen forms in groundwater, glaciers, oceans, lakes and streams. Saline water account for 97.5% of this amount. Fresh water accounts for only 2.5%. Of this fresh water 68.7% is in the "form of ice and permanent snow cover in the Arctic, the Antarctic, and in the mountainous regions.

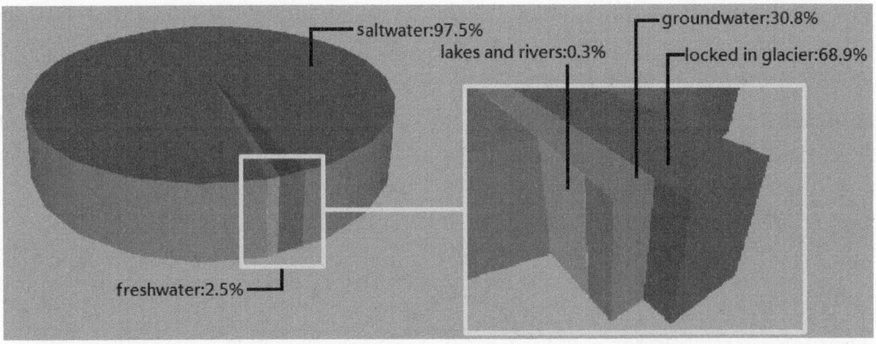

Figure: *The hydrosphere in physical geography describes the combined mass of water found on, under, and over the surface of a planet.*

Next, 29.9% exists as fresh groundwaters. Only 0.26% of the total amount of fresh waters on the Earth is concentrated in lakes, reservoirs and river systems where they are most easily accessible for our economic needs and absolutely vital for water ecosystems." The total mass of the Earth's hydrosphere is about 1.4×10^{18} tonnes, which is about 0.023% of the Earth's total mass. About 20×10^{12} tonnes of this is in the Earth's atmosphere (the volume of one tonne of water is approximately 1 cubic metre). Approximately 75% of the Earth's surface, an area of some 361 million square kilometres (139.5 million square miles), is covered by ocean. The average salinity of the Earth's oceans is about 35 grams of salt per kilogram of sea water (3.5%).

Water Cycle

The hydrological cycle transfers water from one state or reservoir to another. Reservoirs include atmospheric moisture (snow, rain and clouds) oceans, rivers, lakes, groundwater, subterranean aquifers, polar icecaps and saturated soil. Solar energy, in the form of heat and light (insolation), and gravity cause the transfer from one state to another over periods from hours to thousands of years. Most evaporation comes from the oceans and is returned to the earth as snow or rain (page 27). Sublimation refers to evaporation from snow and ice. Transpiration refers to the expiration of water through the minute pores or stomata of trees. Evapotranspiration is the term used by hydrologists in reference to the three processes together, transpiration, sublimation and evaporation.

In an award-winning book titled 'Water'by- [Marq de Villiers] described the hydrosphere as a closed system in which water exists. The hydrosphere is intricate, complex, interdependent, all-pervading and stable and "seems purpose-built for regulating life (de Villiers 2003:26)." De Villiers claimed that, "On earth, the total amount of water has almost certainly not changed since geological times: what we had then we still have. Water can be polluted, abused, and misused but it is neither created nor destroyed, it only migrates. There is no evidence that water vapour escapes into space (page 26)."

"Every year the turnover of water on Earth involves 577,000 km3 of water. This is water that evaporates from the oceanic surface (502,800 km3) and from land (74,200 km3). The same amount of water falls as atmospheric precipitation, 458,000 km3 on the ocean and 119,000 km3 on land. The difference between precipitation and evaporation from the land surface (119,000 - 74,200 = 44,800 km3/year) represents the total runoff of the Earth's rivers (42,700 km3/year) and direct groundwater runoff to the ocean (2100 km3/year). These are the principal sources of fresh water to support life necessities and man's economic activities."

Water is a basic necessity of life. Since 2/3 of the Earth is covered by water, the Earth is also called the blue planet and the watery planet. Hydrosphere plays an important role in the existence of the atmosphere in its present form. Oceans are important in this regard. When the Earth was formed it had only a very thin atmosphere rich in hydrogen and helium similar to the present atmosphere of Mercury. Later the gases hydrogen and helium were expelled from the atmosphere. The gases and water vapour released at the time of the cooling of the Earth helped in the formation of the atmosphere into its present form. Various types of gases and water vapour released by the volcanoes also reached the atmosphere.

When the Earth cooled due to continuous rains the water vapour also condensed and fell as rain. As carbon dioxide in the atmosphere started to dissolve in rain water, atmospheric temperature decreased considerably. This helped the water vapour to condense fast and fall as rain. It is this rain water that filled the depressions on the Earth's surface to result in the formations of the oceans.

It is estimated that these oceans formed about 4000 million years ago. It is in the oceans that the first life forms took shape. They did not breathe oxygen. Later, when cyanobacteria evolved, the process of conversion of carbon dioxide into food and oxygen began. As a result of the Earth's atmospheric composition became distinctly different from that of the other planets. It is this atmosphere of the Earth that is the basis for the survival of life on the Earth.

Recharging Reservoirs

According to Igor A. Shiklomanov, it takes 2500 years for the complete recharge and replenishment of oceanic waters, 10,000 years for permafrost and ice, 1500 years for deep groundwater and mountainous glaciers, 17 years in lakes and 16 days in rivers.

Specific Fresh Water Availability

"Specific water availability is the residual (after use) per capita quantity of fresh water." Fresh water resources are unevenly distributed in terms of space and time and can go from floods to water shortages within months in the same area.

In 1998 76% of the total population had a specific water availability of less than 5.0 thousand m3 per year per capita. Already by 1998, 35% of the global population suffered "very low or catastrophically low water supplies" and Shiklomanov predicted that the situation would deteriorate in the twenty-first century with "most of the Earth's population will be living under the conditions of low or catastrophically low water supply" by 2025.

Sunlight

Sunlight is a portion of the electromagnetic radiation given off by the Sun, in particular infrared, visible, and ultraviolet light. On Earth, sunlight is filtered through the Earth's atmosphere, and is obvious as daylight when the Sun is above the horizon. When the direct solar radiation is not blocked by clouds, it is experienced as sunshine, a combination of bright light and radiant heat. When it is blocked by the clouds or reflects off other objects, it is experienced as diffused light.

The World Meteorological Organization uses the term "sunshine duration" to mean the cumulative time during which an area receives direct irradiance from the Sun of at least 120 watts per square metre.

The ultraviolet B component of sunlight on the skin is an effective source of vitamin D3 (cholecalciferol) from serum cholesterol.

Summary

Sunlight may be recorded using a sunshine recorder, pyranometre, or pyrheliometre.

Sunlight takes about 8.3 minutes to reach the Earth from the surface of the Sun. A photon starting at the centre of the sun and changing direction every time it encounters a charged particle would take between 10 000 and 170 000 years to get to the surface.

The total amount of energy received at ground level from the sun at the zenith depends on the distance to the sun and thus on the time of year. It is about 3.3% higher than average in January and 3.3% lower in July. If the extraterrestrial solar radiation is 1367 watts per square metre (the value when the earth-sun distance is 1 astronomical unit) then the direct sunlight at the earth's surface when the sun is at zenith is about 1050 W/m^2, but the total amount (direct and indirect from the atmosphere) hitting the ground is around 1120 W/m^2. In terms of energy, sunlight at the earth's surface is around 52 or 55 percent infrared (above 700 nm), 43 or 42 percent visible (400 to 700 nm), and 5 or 3 percent ultraviolet (below 400 nm). At the top of the atmosphere sunlight is about 30% more intense, having about 8% ultraviolet (UV), with most of the extra UV consisting of biologically-damaging shortwave ultraviolet.

Direct sunlight has a luminous efficacy of about 93 lumens per watt of radiant flux, higher than most artificial lighting, including fluorescent. Multiplying the figure of 1050 watts per square metre by 93 lumens per watt indicates that bright sunlight provides an illuminance of approximately 98 000 lux (lumens per square metre) on a perpendicular surface at sea level. The illumination of a horizontal surface will be considerably less than this if the sun is not very high in the sky. Averaged over a day, the highest amount of sunlight on a horizontal surface occurs in January at the South Pole.

Sunlight is a key factor in photosynthesis, the process used by plants and other autotrophic organisms to convert light energy, normally from the sun, into chemical energy that can be used to fuel the organisms' activities.

Composition and Power

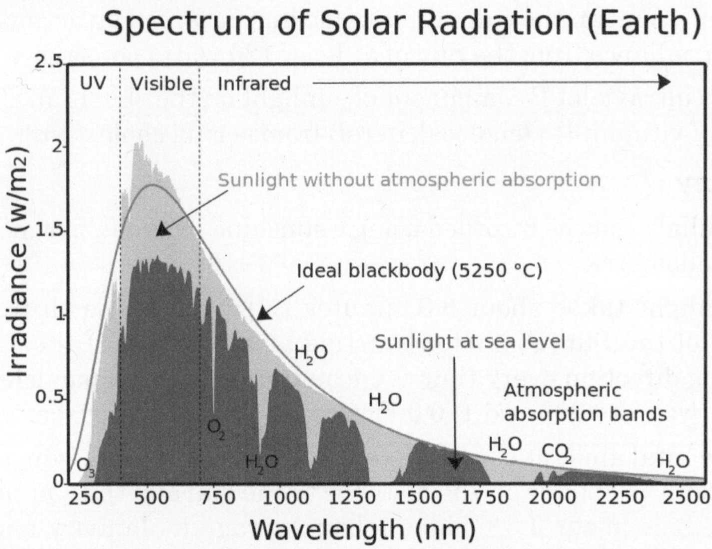

Figure: *Solar irradiance spectrum above atmosphere and at surface. Extreme UV and X-rays are produced (at left of wavelength range shown) but comprise very small amounts of the Sun's total output power.*

The spectrum of the Sun's solar radiation is close to that of a black body with a temperature of about 5,800 K. The Sun emits EM radiation across most of the electromagnetic spectrum. Although the Sun produces Gamma rays as a result of the nuclear fusion process, these super-high-energy photons are converted by internal absorption and thermalization to lower-energy photons before they reach the Sun's surface and are emitted out into space. As a result, the Sun does not emit gamma rays, from this process, although it does produce gamma rays from solar flares. The Sun does also emit X-rays, ultraviolet, visible light, infrared, and even radio waves; the only direct signature of the nuclear process is the emission of neutrinos.

Although the solar corona is a source of extreme ultraviolet and X-ray radiation, these rays make up only a very small amount of the power output of the Sun. The spectrum of nearly all solar electromagnetic radiation striking the Earth's atmosphere spans a range of 100 nm to about 1 mm (1,000,000 nm). This band of significant radiation power can be divided into five regions in increasing order of wavelengths:

- Ultraviolet C or (UVC) range, which spans a range of 100 to 280 nm. The term *ultraviolet* refers to the fact that the radiation is at higher frequency than violet light (and, hence, also invisible to the human eye). Owing to absorption by the atmosphere very little reaches the Earth's surface. This spectrum of radiation has germicidal properties, and is used in germicidal lamps.

- Ultraviolet B or (UVB) range spans 280 to 315 nm. It is also greatly absorbed by the atmosphere, and along with UVC is responsible for the photochemical reaction leading to the production of the ozone layer. It directly damages DNA and causes sunburn.
- Ultraviolet A or (UVA) spans 315 to 400 nm. This band was once held to be less damaging to DNA, and hence is used in cosmetic artificial sun tanning (tanning booths and tanning beds) and PUVA therapy for psoriasis. However, UVA is now known to cause significant damage to DNA via indirect routes (formation of free radicals and reactive oxygen species), and is able to cause cancer.
- Visible range or light spans 380 to 780 nm. As the name suggests, it is this range that is visible to the naked eye. It is also the strongest output range of the sun's total irradiance spectrum.
- Infrared range that spans 700 nm to 1,000,000 nm (1 mm). It is responsible for an important part of the electromagnetic radiation that reaches the Earth. It is also divided into three types on the basis of wavelength:
    - o Infrared-A: 700 nm to 1,400 nm
    - o Infrared-B: 1,400 nm to 3,000 nm
    - o Infrared-C: 3,000 nm to 1 mm.

Published Tables

Tables of direct solar radiation on various slopes from 0 to 60 degrees North Latitude, in calories per square centimetre were issued in 1972 are published by Pacific Northwest Forest and Range Experiment Station | Forest Service, U.S. Department of Agriculture, Portland, Oregon, USA.

Calculation

To calculate the amount of sunlight reaching the ground, both the elliptical orbit of the Earth and the attenuation by the Earth's atmosphere have to be taken into account. The extraterrestrial solar illuminance (E_{ext}), corrected for the elliptical orbit by using the day number of the year (dn), is given to a good approximation by

$$E_{ext} = E_{sc} \cdot \left(1 + 0.033412 \cdot \cos\left(2\pi \frac{dn - 3}{365}\right)\right),$$

where dn=1 on January 1; dn=2 on January 2; dn=32 on February 1, etc. In this formula dn-3 is used, because in modern times Earth's perihelion, the closest approach to the Sun and, therefore, the maximum E_{ext} occurs around January 3 each year.

The value of 0.033412 is determined knowing that the ratio between the perihelion (0.98328989 AU) squared and the aphelion (1.01671033 AU) squared should be approximately 0.935338.

The solar illuminance constant (E_{sc}), is equal to 128×10³ lx. The direct normal illuminance (E_{dn}), corrected for the attenuating effects of the atmosphere is given by:

$$E_{dn} = E_{ext}\, e^{-cm},$$

where c is the atmospheric extinction and m is the relative optical airmass. The atmospheric extinction brings the number of lux down to around 100 000.

Total (TSI) and Spectral Solar Irradiance (SSI) upon Earth

Total Solar Irradiance (TSI) – the amount of solar radiation received at the top of the Earth's atmosphere – has been measured since 1978 by series of overlapping NASA and ESA satellite experiments to be 1.361 kilo' watts per square metre (kW/m²). TSI observations are continuing today with the ACRIMSAT/ACRIM3, SOHO/VIRGO and SORCE/TIM satellite experiments.

Variation of TSI has been discovered on many timescales including the solar magnetic cycle and many shorter periodic cycles. TSI provides the energy that drives the Earth's climate, so continuation of the TSI time series database is critical to understanding the role of solar variability in climate change.

Spectral Solar Irradiance (SSI) - the spectral distribution of the TSI - has been monitored since 2003 by the SORCE Spectral Irradiance Monitor (SIM). It has been found that SSI at UV (ultraviolet) wavelength corresponds in a less clear, and probably more complicated fashion, with Earth's climate responses than earlier assumed, fuelling broad avenues of new research in "the connection of the Sun and stratosphere, troposphere, biosphere, ocean, and Earth's climate".

Intensity in the Solar System

Different bodies of the Solar System receive light of an intensity inversely proportional to the square of their distance from Sun. A rough table comparing the amount of solar radiation received by each planet in the Solar System follows (from data in:

Planet or dwarf planet	***distance (AU)***		***Solar radiation (W/m²)***	
	Perihelion	***Aphelion***	***maximum***	***minimum***
Mercury	0.3075	0.4667	14,446	6,272
Venus	0.7184	0.7282	2,647	2,576
Earth	0.9833	1.017	1,413	1,321
Mars	1.382	1.666	715	492
Jupiter	4.950	5.458	55.8	45.9
Saturn	9.048	10.12	16.7	13.4
Uranus	18.38	20.08	4.04	3.39
Neptune	29.77	30.44	1.54	1.47
Pluto	29.66	48.87	1.55	0.57

The actual brightness of sunlight that would be observed at the surface depends also on the presence and composition of an atmosphere. For example, Venus's thick atmosphere reflects more than 60% of the solar light it receives. The actual illumination of the surface is about 14,000 lux, comparable to that on Earth "in the daytime with overcast clouds".

Sunlight on Mars would be more or less like daylight on Earth wearing sunglasses, and, as can be seen in the pictures taken by the rovers, there is enough diffuse sky radiation that shadows would not seem particularly dark. Thus, it would give perceptions and "feel" very much like Earth daylight.

For comparison purposes, sunlight on Saturn is slightly brighter than Earth sunlight at the average sunset or sunrise. Even on Pluto the sunlight would still be bright enough to almost match the average living room. To see sunlight as dim as full moonlight on Earth, a distance of about 500 AU (~69 light-hours) is needed; there are only a handful of objects in the Solar System known to orbit farther than such a distance, among them 90377 Sedna and (87269) 2000 OO67.

Surface Illumination

The spectrum of surface illumination depends upon solar elevation due to atmospheric effects, with the blue spectral component dominating during twilight before and after sunrise and sunset, respectively, and red dominating during sunrise and sunset. These effects are apparent in natural light photography where the principal source of illumination is sunlight as mediated by the atmosphere.

While the colour of the sky is usually determined by Rayleigh scattering, an exception occurs at sunset and twilight. "Preferential absorption of sunlight by ozone over long horizon paths gives the zenith sky its blueness when the sun is near the horizon".

Spectral Composition of Surface Illumination

The Sun's electromagnetic radiation which illuminates the earth's surface is predominantly light that falls within the range of wavelengths to which the visual systems of the animals that inhabit the earth's surface are sensitive. The Sun may therefore be said to illuminate, which is a measure of the light within a specific sensitivity range. Many animals (including humans) have a sensitivity range of approximately 400–700 nm, and given optimal conditions the absorption and scattering by the earth's atmosphere produces illumination that approximates an equal-energy illuminant for most of this range. The useful range for colour vision in humans, for example, is approximately 450–650 nm. Aside from effects that arise at sunset and sunrise, the spectral composition changes primarily in respect to how directly sunlight is able to illuminate. When illumination is indirect, Rayleigh scattering in the upper atmosphere will lead blue wavelengths to dominate. Water vapour in the lower atmosphere produces further scattering and ozone, dust and water particles will also absorb selective wavelengths.

Climate Effects

On Earth, solar radiation is obvious as daylight when the sun is above the horizon. This is during daytime, and also in summer near the poles at night, but not at all in winter near the poles. When the direct radiation is not blocked by clouds, it is experienced as *sunshine*, combining the perception of bright white light (sunlight in the strict sense) and warming. The warming on the body, the ground, and other objects depends on the absorption (electromagnetic radiation) of the electromagnetic radiation in the form of heat. The amount of radiation intercepted by a planetary body varies inversely with the square of the distance between the star and the planet. The Earth's orbit and obliquity change with time (over thousands of years), sometimes forming a nearly perfect circle, and at other times stretching out to an orbital eccentricity of 5% (currently 1.67%). As the orbit changes, the total insolation over a year remains almost constant due to Kepler's second law,

$$\frac{2A}{r^2}dt = d\theta,$$

where A is the "areal velocity" invariant. That is, the integration over the orbital period (also invariant) is a constant.

$$\int_0^T \frac{2A}{r^2}dt = \int_0^{2\pi} d\theta = \text{constant}.$$

If we assume the solar radiation power P as a constant over time and the solar irradiation given by the inverse-square law, we obtain also the average insolation as a constant.

But the seasonal and latitudinal distribution and intensity of solar radiation received at the Earth's surface also varies. For example, at latitudes of 65 degrees, the change in solar energy in summer and winter can vary by more than 25% as a result of the Earth's orbital variation. Because changes in winter and summer tend to offset, the change in the annual average insolation at any given location is near zero, but the redistribution of energy between summer and winter does strongly affect the intensity of seasonal cycles. Such changes associated with the redistribution of solar energy are considered a likely cause for the coming and going of recent ice ages.

Past Variations in Solar Irradiance

Space-based observations of solar irradiance started in 1978. These measurements show that the solar constant is not constant. It varies on many periodic cycles including the 11-year sunspot solar cycle. When going further back in time, one has to rely on irradiance reconstructions, using sunspots for the past 400 years or cosmogenic radionuclides for going back 10,000 years. Such reconstructions have been done. These studies show that solar irradiance does vary with distinct periodicities such as: 11 years (Schwabe), 88 years (Gleisberg cycle), 208 years (DeVries cycle) and 1,000 years (Eddy cycle).

Life on Earth

The existence of nearly all life on Earth is fuelled by light from the sun. Most autotrophs, such as plants, use the energy of sunlight, combined with carbon dioxide and water, to produce simple sugars—a process known as photosynthesis. These sugars are then used as building-blocks and in other synthetic pathways that allow the organism to grow.

Heterotrophs, such as animals, use light from the sun indirectly by consuming the products of autotrophs, either by consuming autotrophs, by consuming their products, or by consuming other heterotrophs. The sugars and other molecular components produced by the autotrophs are then broken down, releasing stored solar energy, and giving the heterotroph the energy required for survival. This process is known as cellular respiration.

In prehistory, humans began to further extend this process by putting plant and animal materials to other uses. They used animal skins for warmth, for example, or wooden weapons to hunt. These skills allowed humans to harvest more of the sunlight than was possible through glycolysis alone, and human population began to grow. During the Neolithic Revolution, the domestication of plants and animals further increased human access to solar energy. Fields devoted to crops were enriched by inedible plant matter, providing sugars and nutrients for future harvests. Animals that had previously provided humans with only meat and tools once they were killed were now used for labour throughout their lives, fuelled by grasses inedible to humans.

The more recent discoveries of coal, petroleum and natural gas are modern extensions of this trend. These fossil fuels are the remnants of ancient plant and animal matter, formed using energy from sunlight and then trapped within the earth for millions of years. Because the stored energy in these fossil fuels has accumulated over many millions of years, they have allowed modern humans to massively increase the production and consumption of primary energy. As the amount of fossil fuel is large but finite, this cannot continue indefinitely, and various theories exist as to what will follow this stage of human civilization (e.g., alternative fuels, Malthusian catastrophe, new urbanism, peak oil).

Solar Radiation and its Terms

Agriculture is the exploitation of solar energy under adequate supply of nutrients and water by maintaining plant growth. So it is but natural that any efforts of thoroughly understanding of solar radiation will be immense use for its fullest exploitation by the crop plants in terms of their growth and yield.

The sun is the primary source of energy. Supplying about 99.9% out of total energy available at the earth surface. The temperature of the Sun is 6000 K and gives out energy about 5.6 x10 27 cal per minute. The Sun radiates its energy in the form of wave lengths from 0.15 to 4.0 u and are generally called as short wave lengths. On contrary after absorption of solar energy, earth emits its energy between 4 to 100 u and is categorized as long wave length.

There are three methods of transfer of heat or energy that means there are three different ways by which heat can flow from one point to another are:

1. Conduction
2. Convection
3. Radiation.

For conduction and convection of heat, material medium is necessary. But for radiation material medium is not necessary, because radiation takes place in the form of Electro magnetic waves.

The ultimate source of all the energy for physical and biological processes occurring on the earth is radiation received from the sun that is why it is commonly called solar radiation.

Some Terms (Definitions)

Radiation: The transfer of heat energy in the form of electro magnetic waves with the speed of light is known as radiation (Light speed is 3x105 km/second).

Solar insulation: The heat energy received from the Sun is known as solar insulation.

Radiant flux density: It is defined as the amount of energy received on a unit surface in a unit time. (In Meteorology we commonly use cal cm -2 min or largely min -1 as the unit of radiant flux density)

Emissive: It is defined as the ratio of the emittance of a given surface at a specified wave length and temperature to the emittance of an ideal black body at the same wave length and temperature.

Absorptive: It is defined as the ratio of the amount of radiant energy absorbed to the total amount incident upon the substance.

Reflectivity: Is defined as the ratio of the radiant energy reflected to the total that is incident upon the surface.

Transmmissivity: Is defined as the ratio of the transmitted radiation to the total radiation incident upon the medium.

Short wave radiation (SW): Radiation with wave length range 0.15 to 0.76 u is known as short wave radiation.

Long wave radiation: Radiation with wave length range 0.76 to 100 u is defined as long wave radiation.

Chapter 7

Laws of Radiation

The radiation reaching to the earth surface from the sun, atmosphere and from earth to atmosphere, space follows certain physical laws known as radiation laws. They are

Plank's Law

The Electro magnetic radiation consists of a stream or a flow of particles or quanta. Each quantum having energy content

E=hv. Where, h=Plank's constant (6.625x1027 ergs sec-1)

V=frequency of Electro-magnetic length.

Greater the frequency (i.e. shorter the wave length) greater is the energy content of the quantum.

Stefan–Boltzmann Law

The Stefan–Boltzmann law, also known as Stefan's law, describes the power radiated from a black body in terms of its temperature.

Specifically, the Stefan–Boltzmann law states that the total energy radiated per unit surface area of a black body across all wavelengths per unit time (also known as the black-body *radiant exitance* or *emissive power*), $j^{\star}$, is directly proportional to the fourth power of the black body's thermodynamic temperature *T*:

$$j^{\star} = \sigma T^4.$$

The constant of proportionality σ, called the Stefan–Boltzmann constant or Stefan's constant, derives from other known constants of nature.

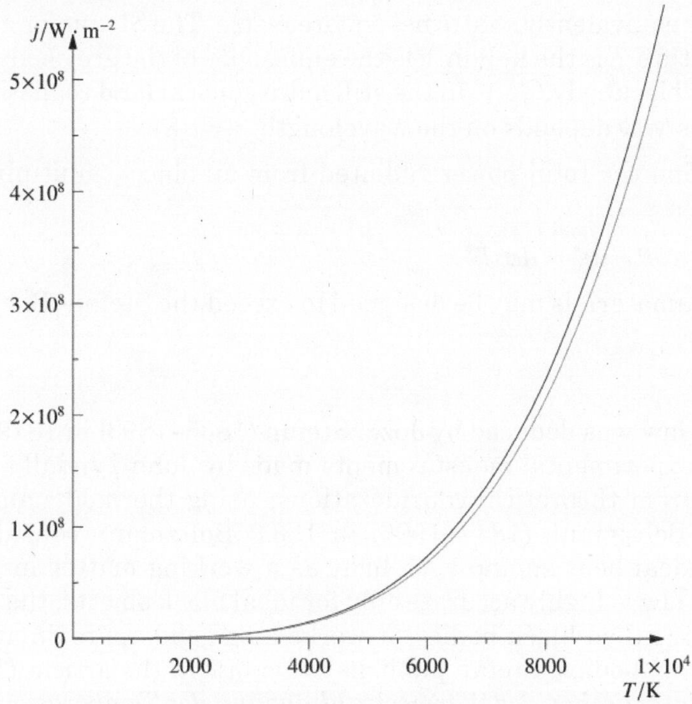

Figure: *Graph of a function of total emitted energy of a black body $j^\star$ proportional to its thermodynamic temperature T. In blue is a total energy according to the Wien approximation,* $j_W^\star = j^\star / \zeta(4) \approx 0.924\sigma T^4$

The value of the constant is

$$\sigma = \frac{2\pi^5 k^4}{15c^2h^3} = 5.670373\times10^{-8}\ \mathrm{Js^{-1}m^{-2}K^{-4}},$$

where k is the Boltzmann constant, h is Planck's constant, and c is the speed of light in a vacuum. Thus at 100 K the energy flux is 5.67 W/m^2, at 1000 K 56,700 W/m^2, etc. The radiance (watts per square metre per steradian) is given by

$$L = \frac{j^\star}{\pi} = \frac{\sigma}{\pi}T^4.$$

A body that does not absorb all incident radiation (sometimes known as a grey body) emits less total energy than a black body and is characterized by an emissivity, $\varepsilon < 1$:

$$j^\star = \varepsilon\sigma T^4.$$

The irradiance $j^\star$ has dimensions of energy flux (energy per time per area), and the SI units of measure are joules per second per square metre, or equivalently, watts per square metre. The SI unit for absolute temperature T is the kelvin. ε is the emissivity of the grey body; if it is a perfect blackbody, $\varepsilon = 1$. In the still more general (and realistic) case, the emissivity depends on the wavelength, $\varepsilon = \varepsilon(\lambda)$.

To find the total power radiated from an object, multiply by its surface area, A:

$$P = Aj^{\star} = A\varepsilon\sigma T^4.$$

Metamaterials may be designed to exceed the Stefan–Boltzmann law.

History

The law was deduced by Jožef Stefan (1835–1893) in 1879 on the basis of experimental measurements made by John Tyndall and was derived from theoretical considerations, using thermodynamics, by Ludwig Boltzmann (1844–1906) in 1884. Boltzmann considered a certain ideal heat engine with light as a working matter instead of gas. The law is highly accurate only for ideal black objects, the perfect radiators, called black bodies; it works as a good approximation for most "grey" bodies. Stefan published this law in the article *Über die Beziehung zwischen der Wärmestrahlung und der Temperatur* (*On the relationship between thermal radiation and temperature*) in the *Bulletins from the sessions* of the Vienna Academy of Sciences.

Examples

Temperature of the Sun: With his law Stefan also determined the temperature of the Sun's surface. He learned from the data of Charles Soret (1854–1904) that the energy flux density from the Sun is 29 times greater than the energy flux density of a certain warmed metal lamella (a thin plate). A round lamella was placed at such a distance from the measuring device that it would be seen at the same angle as the Sun. Soret estimated the temperature of the lamella to be approximately 1900 °C to 2000 °C. Stefan surmised that S! of the energy flux from the Sun is absorbed by the Earth's atmosphere, so he took for the correct Sun's energy flux a value 3/2 times greater, namely 29 × 3/2 = 43.5.

Precise measurements of atmospheric absorption were not made until 1888 and 1904. The temperature Stefan obtained was a median value of previous ones, 1950 °C and the absolute thermodynamic one 2200 K. As $2.57^4 = 43.5$, it follows from the law that the temperature of the Sun is 2.57 times greater than the temperature of the lamella, so Stefan got a value of 5430 °C or 5700 K (the modern value is 5778 K).

This was the first sensible value for the temperature of the Sun. Before this, values ranging from as low as 1800 °C to as high as 13,000,000 °C were claimed. The lower value of 1800 °C was determined by Claude Servais Mathias Pouillet (1790–1868) in 1838 using the Dulong-Petit law. Pouillet also took just half the value of the Sun's correct energy flux.

Temperature of Stars

The temperature of stars other than the Sun can be approximated using a similar means by treating the emitted energy as a black body radiation. So:

$$L = 4\pi R^2 \sigma T_e^4$$

where L is the luminosity, σ is the Stefan–Boltzmann constant, R is the stellar radius and T is the effective temperature. This same formula can be used to compute the approximate radius of a main sequence star relative to the sun:

$$\frac{R}{R_\odot} \approx \left(\frac{T_\odot}{T}\right)^2 \cdot \sqrt{\frac{L}{L_\odot}}$$

where $R_\odot$, is the solar radius, and so forth.

With the Stefan–Boltzmann law, astronomers can easily infer the radii of stars. The law is also met in the thermodynamics of black holes in so-called Hawking radiation.

Temperature of the Earth

Similarly we can calculate the effective temperature of the Earth T_E by equating the energy received from the Sun and the energy radiated by the Earth, under the black-body approximation. The amount of power, E_S, emitted by the Sun is given by:

$$E_S = 4\pi r_S^2 \sigma T_S^4$$

At Earth, this energy is passing through a sphere with a radius of a_0, the distance between the Earth and the Sun, and the energy passing through each square metre of the sphere is given by

$$E_{a_0} = \frac{E_S}{4\pi a_0^2}$$

The Earth has a radius of r_E, and therefore has a cross-section of πr_E^2. The amount of solar power absorbed by the Earth is thus given by:

$$E_{abs} = \pi r_E^2 \times E_{a_0} :$$

Assuming the exchange is in a steady state, the amount of energy emitted by Earth must equal the amount absorbed, and so:

$$\begin{aligned} 4\pi r_E^2 \sigma T_E^4 &= \pi r_E^2 \times E_{a_0} \\ &= \pi r_E^2 \times \frac{4\pi r_S^2 \sigma T_S^4}{4\pi a_0^2} \end{aligned}$$

T_E can then be found:

$$\begin{aligned} T_E^4 &= \frac{r_S^2 T_S^4}{4a_0^2} \\ T_E &= T_S \times \sqrt{\frac{r_S}{2a_0}} \\ &= 5780\ \text{K} \times \sqrt{\frac{696 \times 10^6\ \text{m}}{2 \times 149.598 \times 10^9\ \text{m}}} \\ &\approx 279\ \text{K} \end{aligned}$$

where T_S is the temperature of the Sun, r_S the radius of the Sun, and a_0 is the distance between the Earth and the Sun. This gives an effective temperature of 6 °C on the surface of the Earth, assuming that it perfectly absorbs all emission falling on it and has no atmosphere.

The Earth has an albedo of 0.3, meaning that 30% of the solar radiation that hits the planet gets scattered back into space without absorption. The effect of albedo on temperature can be approximated by assuming that the energy absorbed is multiplied by 0.7, but that the planet still radiates as a black body (the latter by definition of effective temperature, which is what we are calculating). This approximation reduces the temperature by a factor of $0.7^{1/4}$, giving 255 K (–18 °C).

However, long-wave radiation from the surface of the earth is partially absorbed and re-radiated back down by greenhouse gases, namely water vapour, carbon dioxide and methane. Since the emissivity with greenhouse effect (weighted more in the longer wavelengths where the Earth radiates) is reduced more than the absorptivity (weighted more in the shorter wavelengths of the Sun's radiation) is reduced, the equilibrium temperature is higher than the simple black-body calculation estimates.

As a result, the Earth's actual average surface temperature is about 288 K (15 °C), which is higher than the 255 K effective temperature, and even higher than the 279 K temperature that a black body would have.

Origination

Thermodynamic Derivation of the Energy Density

The fact that the energy density of the box containing radiation is proportional to T^4 can be derived using thermodynamics. It follows from classical electrodynamics that the radiation pressure ρ is related to the internal energy density u:

$$p = \frac{u}{3}.$$

From the fundamental thermodynamic relation

$$dU = TdS - pdV,$$

we obtain the following expression, after dividing by dV and fixing T:

$$\left(\frac{\partial U}{\partial V}\right)_T = T\left(\frac{\partial S}{\partial V}\right)_T - p = T\left(\frac{\partial p}{\partial T}\right)_V - p.$$

The last equality comes from the following Maxwell relation:

$$\left(\frac{\partial S}{\partial V}\right)_T = \left(\frac{\partial p}{\partial T}\right)_V.$$

From the definition of energy density it follows that

$$u = \frac{U}{V}$$

and

$$\left(\frac{\partial U}{\partial V}\right)_T = u.$$

Now, the equality

$$\left(\frac{\partial U}{\partial V}\right)_T = T\left(\frac{\partial p}{\partial T}\right)_V - p,$$

after substitution of $\left(\frac{\partial U}{\partial V}\right)_T$ and p for the corresponding expressions, can be written as

$$u = \frac{T}{3}\left(\frac{\partial u}{\partial T}\right)_V - \frac{u}{3}.$$

Since the partial derivative $\left(\frac{\partial u}{\partial T}\right)_V$ can be expressed as a relationship between only u and T (if one isolates it on one side of the equality), the partial derivative can be replaced by the ordinary derivative. After separating the differentials the equality becomes

$$\frac{du}{4u} = \frac{dT}{T},$$

which leads immediately to $u = AT^4$, with A as some constant of integration.

Stefan–Boltzmann's Law in n-dimensional Space

It can be shown that the radiation pressure in n-dimensional space is given by

$$P = \frac{u}{n}$$

So in n-dimensional space,

$$TdS = (n+1)PdV + nVdP$$

So,

$$\frac{1}{P}\frac{dP}{dT} = \frac{(n+1)}{T}$$

yielding

$$P \propto T^{n+1}$$

or

$$u \propto T^{n+1}$$

implying

$$\frac{dQ}{dt} \propto T^{n+1}$$

The same result is obtained as the integral over frequency of Planck's law for n-dimensional space, albeit with a different value for the Stefan-Boltzmann constant at each dimension. In general the constant is

$$\sigma = \frac{1}{p(n)} \frac{\pi^{\frac{n}{2}}}{\Gamma(1+\frac{n}{2})} \frac{1}{c^{n-1}} \frac{n(n-1)}{h^n} k^{(n+1)} \Gamma(n+1)\zeta(n+1)$$

where $\zeta(x)$ is Riemann's zeta function and $p(n)$ is a certain function of n, with $p(3) = 4$.

Derivation from Planck's Law

The law can be derived by considering a small flat black body surface radiating out into a half-sphere. This derivation uses spherical coordinates, with φ as the zenith angle and θ as the azimuthal angle; and the small flat blackbody surface lies on the xy-plane, where $\varphi = \pi/_2$.

The intensity of the light emitted from the blackbody surface is given by Planck's law :

$$I(\nu,T) = \frac{2h\nu^3}{c^2}\frac{1}{e^{\frac{h\nu}{kT}}-1}.$$

where

- $I(\nu,T)$ is the amount of energy per unit surface area per unit time per unit solid angle emitted at a frequency ν by a black body at temperature T.
- h is Planck's constant
- c is the speed of light, and
- k is Boltzmann's constant.

The quantity $I(\nu,T)A\,d\nu\,d\Omega$ is the power radiated by a surface of area A through a solid angle $d\Omega$ in the frequency range between ν and $\nu + d\nu$.

The Stefan–Boltzmann law gives the power emitted per unit area of the emitting body,

$$\frac{P}{A} = \int_0^\infty I(\nu,T)d\nu \int d\Omega$$

To derive the Stefan–Boltzmann law, we must integrate Ω over the half-sphere and integrate ν from 0 to ∞. Furthermore, because black bodies are *Lambertian* (i.e. they obey Lambert's cosine law), the intensity observed along the sphere will be the actual intensity times the cosine of the zenith angle φ, and in spherical coordinates, $d\Omega = \sin(\varphi)\, d\varphi\, d\theta$.

$$\begin{aligned}\frac{P}{A} &= \int_0^\infty I(\nu,T)d\nu \int_0^{2\pi} d\theta \int_0^{\pi/2} \cos\phi \sin\phi\, d\phi \\ &= \pi \int_0^\infty I(\nu,T)d\nu\end{aligned}$$

Then we plug in for I:

$$\frac{P}{A} = \frac{2\pi h}{c^2}\int_0^\infty \frac{\nu^3}{e^{\frac{h\nu}{kT}}-1}d\nu$$

To do this integral, do a substitution,

$$u = \frac{h\nu}{kT}$$

$$du = \frac{h}{kT} d\nu$$

which gives:

$$\frac{P}{A} = \frac{2\pi h}{c^2}\left(\frac{kT}{h}\right)^4 \int_0^\infty \frac{u^3}{e^u - 1} du.$$

The integral on the right can be done in a number of ways, its answer is $\frac{\pi^4}{15}$, giving the result that, for a perfect blackbody surface:

$$j^\star = \sigma T^4 \ , \sigma = \frac{2\pi^5 k^4}{15c^2 h^3} = \frac{\pi^2 k^4}{60\hbar^3 c^2}.$$

Finally, this proof started out only considering a small flat surface. However, any differentiable surface can be approximated by a bunch of small flat surfaces. So long as the geometry of the surface does not cause the blackbody to reabsorb its own radiation, the total energy radiated is just the sum of the energies radiated by each surface; and the total surface area is just the sum of the areas of each surface—so this law holds for all convex blackbodies, too, so long as the surface has the same temperature throughout. The law extends to radiation from non-convex bodies by using the fact that the convex hull of a black body radiates as though it were itself a black body.

Appendix

In one of the above derivations, the following integral appeared:

$$J = \int_0^\infty \frac{x^3}{\exp(x) - 1} dx = \Gamma(4)\mathrm{Li}_4(1) = 6\mathrm{Li}_4(1) = 6\zeta(4)$$

where $\mathrm{Li}_s(z)$ is the polylogarithm function and $\zeta(z)$ is the Riemann zeta function. If the polylogarithm function and the Riemann zeta function are not available for calculation, there are a number of ways to do this integration; a simple one is given in the appendix of the Planck's law article. This appendix does the integral by contour integration. Consider the function:

$$f(k) = \int_0^\infty \frac{\sin(kx)}{\exp(x) - 1} dx.$$

Using the Taylor expansion of the sine function, it should be evident that the coefficient of the k^3 term would be exactly $-J/6$. By expanding both sides in powers of k, we see that J is minus 6 times the coefficient of k^3 of the series expansion of $f(k)$. So, if we can find a closed form for $f(k)$, its Taylor expansion will give J.

In turn, sin(x) is the imaginary part of e^{ix}, so we can restate this as:

$$f(k) = \lim_{\varepsilon\to 0} \operatorname{Im} \int_{\varepsilon}^{\infty} \frac{\exp(ikx)}{\exp(x)-1} dx.$$

To evaluate the integral in this equation we consider the contour integral:

$$\oint_{C(\varepsilon,R)} \frac{\exp(ikz)}{\exp(z)-1} dz$$

where $C(\varepsilon,R)$ is the contour from ε to R, then to $R+2\pi i$, then to $\varepsilon+2\pi i$, then we go to the point $2\pi i-\varepsilon i$, avoiding the pole at $2\pi i$ by taking a clockwise quarter circle with radius ε and centre $2\pi i$. From there we go to εi, and finally we return to ε, avoiding the pole at zero by taking a clockwise quarter circle with radius ε and centre zero.

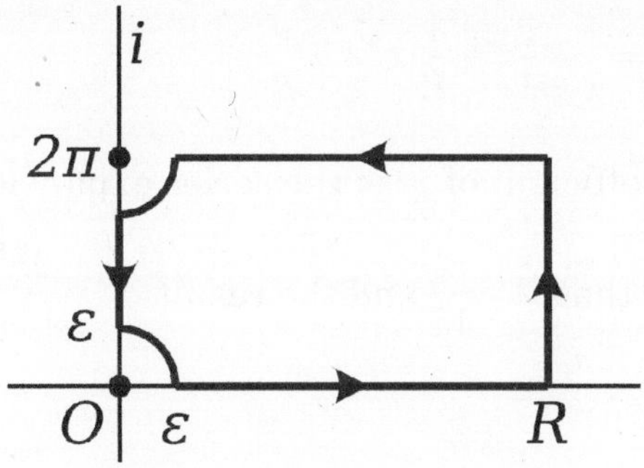

Figure: *Integration contour*

Because there are no poles in the integration contour we have:

$$\oint_{C(\varepsilon,R)} \frac{\exp(ikz)}{\exp(z)-1} dz = 0.$$

We now take the limit $R\to\infty$. In this limit the contribution from the segment from R to $R+2\pi i$ tends to zero. Taking together the integrations over the segments from ε to R and from $R+2\pi i$ to $\varepsilon+2\pi i$ and using the fact that the integrations over clockwise quarter circles withradius ε about simple poles are given up to order ε by minus $\frac{i\pi}{2}$ times the residues at the poles we find:

$$\left[1-\exp(-2\pi k)\right]\int_{\varepsilon}^{\infty} \frac{\exp(ikx)}{\exp(x)-1} dx = i\int_{\varepsilon}^{2\pi-\varepsilon} \frac{\exp(-ky)}{\exp(iy)-1} dy + i\frac{\pi}{2}\left[1+\exp(-2\pi k)\right] + \mathcal{O}(\varepsilon) \qquad (1)$$

The left hand side is the sum of the integral from ε to R and from $R+2\pi i$ to $2\pi i+\varepsilon$. We can rewrite the integrand of the integral on the r.h.s. as follows:

$$\frac{1}{\exp(iy)-1}=\frac{\exp\left(-i\frac{y}{2}\right)}{\exp\left(i\frac{y}{2}\right)-\exp\left(-i\frac{y}{2}\right)}=\frac{1}{2i}\frac{\exp\left(-i\frac{y}{2}\right)}{\sin\left(\frac{y}{2}\right)}$$

If we now take the imaginary part of both sides of Eq. (1) and take the limit $\varepsilon \to 0$ we find:

$$f(k)=-\frac{1}{2k}+\frac{\pi}{2}\coth(\pi k)$$

after using the relation:

$$\coth(x)=\frac{1+\exp(-2x)}{1-\exp(-2x)}.$$

Using that the series expansion of $\coth(x)$ is given by:

$$\coth(x)=\frac{1}{x}+\frac{1}{3}x-\frac{1}{45}x^3+\cdots$$

we see that the coefficient of k^3 of the series expansion of $f(k)$ is $-\frac{\pi^4}{90}$.

This then implies that $J=\frac{\pi^4}{15}$ and the result $j^{\star}=\frac{2\pi^5 k^4}{15h^3c^2}T^4$ follows.

Planck's Law

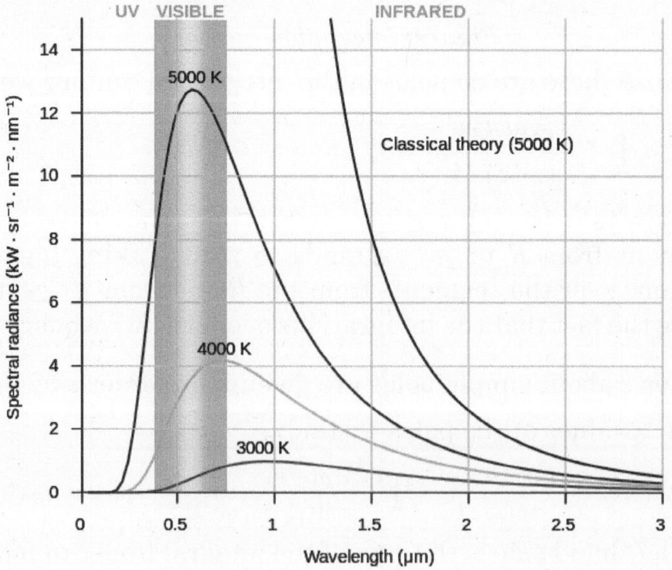

Figure: *Planck's law (coloured curves) accurately described black body radiation and resolved the ultraviolet catastrophe (black curve).*

Planck's law describes the electromagnetic radiation emitted by a black body in thermal equilibrium at a definite temperature. The law is named after Max Planck, who originally proposed it in 1900. It is a pioneering result of modern physics and quantum theory.

The spectral radiance of a body, B_ν, describes the amount of energy it gives off as radiation of different frequencies. It is measured in terms of the power emitted per unit area of the body, per unit solid angle that the radiation is measured over, per unit frequency. Planck showed that the spectral radiance of a body at absolute temperature T is given by

$$B_\nu(\nu,T) = \frac{2h\nu^3}{c^2}\frac{1}{e^{\frac{h\nu}{k_B T}}-1}$$

where k_B the Boltzmann constant, h the Planck constant, and c the speed of light in the medium, whether material or vacuum. The spectral radiance can also be measured per unit wavelength instead of per unit frequency. In this case, it is given by

$$B_\lambda(\lambda,T) = \frac{2hc^2}{\lambda^5}\frac{1}{e^{\frac{hc}{\lambda k_B T}}-1}.$$

The SI units are W·sr^{-1}·m^{-2}·Hz^{-1} of B_ν and W·sr^{-1}·m^{-3} for B_λ. The law may also be expressed in other terms, such as of the number of photons emitted at a certain wavelength, or of the energy density in a volume of radiation.

In the limit of low frequencies (i.e. long wavelengths), Planck's law tends to the Rayleigh–Jeans law, while in the limit of high frequencies (i.e. small wavelengths) it tends to the Wien approximation.

Max Planck developed the law in 1900, originally with only empirically determined constants, and later showed that, expressed as an energy distribution, it is the unique stable distribution for radiation in thermodynamic equilibrium. As an energy distribution, it is one of a family of thermal equilibrium distributions which include the Bose–Einstein distribution, the Fermi–Dirac distribution and the Maxwell–Boltzmann distribution.

Introduction

Every physical body spontaneously and continuously emits electromagnetic radiation. Near thermodynamic equilibrium, the emitted radiation is nearly described by Planck's law. Because of its dependence on temperature, Planck radiation is said to be thermal.

The higher the temperature of a body the more radiation it emits at every wavelength. Planck radiation has a maximum intensity at a specific wavelength that depends on the temperature. For example, at room temperature (~300 K), a body emits thermal radiation that is mostly infrared and invisible.

At higher temperatures the amount of infrared radiation increases and can be felt as heat, and the body glows visibly red. At even higher temperatures, a body is dazzlingly bright yellow or blue-white and emits significant amounts of short wavelength radiation, including ultraviolet and even x-rays. The surface of the sun (~6000 K) emits large amounts of both infrared and ultraviolet radiation; its emission is peaked in the visible spectrum.

In the interior of a physical medium, radiation can be absorbed and emitted by matter. This mediates transfer of energy as heat, and can change the internal energy of the matter, and the occupation numbers of the states of its molecules.

Planck radiation is the greatest amount of radiation that any body at thermal equilibrium can emit from its surface, whatever its chemical composition or surface structure. Passage of radiation across an interface between media can be characterized by the emissivity of the interface, the radiance of the passing radiation divided by the Planck radiance. It is in general dependent on chemical composition and physical structure, on temperature, on the wavelength, on the angle of passage, and on the polarization, of the radiation.

The emissivity of an interface is also known as its transmittance or as its absorbance. The emissivity of a natural interface is always between zero and one. For an interface, the sum of emissivity and reflectivity is one. An ideally perfectly reflecting interface has emissivity zero, reflectivity one. An ideally perfectly transmitting interface has emissivity one, reflectivity zero.

A body that interfaces with another medium with emissivity one, and that absorbs all the radiation incident upon it, is said to be a black body. The surface of a black body can be modelled by a small hole in the wall of a large enclosure which is maintained at a uniform temperature with rigid opaque walls that are not perfectly reflective at any wavelength. At equilibrium, the radiation inside this enclosure follows Planck's law. This radiation is well sampled by the radiation that is emitted at right angles from the hole.

Just as the Maxwell–Boltzmann distribution for thermodynamic equilibrium at a given temperature is the unique maximum entropy energy distribution for a gas of many conserved massive particles, so also is Planck's distribution for a gas of photons, which are not conserved and have zero rest mass. By contrast to a material gas where the masses and number of particles play a role, the spectral radiance, pressure and energy density of a photon gas at equilibrium are entirely determined by the temperature.

If the photon gas is not initially Planckian, the second law of thermodynamics guarantees that interactions (between photons and other particles or even between the photons themselves) will cause the photon energy distribution to change and approach the Planck distribution. In such an approach to thermodynamic equilibrium photons are created or annihilated in the right numbers and with the right energies to fill the cavity with a Planck distribution at the eventual equilibrium temperature.

The quantity $B_\nu(T)$ is the spectral radiance as a function of temperature and frequency. It has units of $W \cdot m^{-2} \cdot sr^{-1} \cdot Hz^{-1}$ in the SI system. An infinitesimal amount of power $B_\nu(T) \cos \theta \, dA \, d\Omega \, d\nu$ is radiated in the direction described by the angle θ from the surface normal from infinitesimal surface area dA into infinitesimal solid angle $d\Omega$ in an infinitesimal frequency band of width $d\nu$ centred on frequency ν. The total power radiated into any solid angle is the integral of $B_\nu(T)$ over those three quantities, and is given by the Stefan–Boltzmann law. The spectral radiance of Planckian radiation from a black body has the same value for every direction and angle of polarization, and so the black body is said to be a Lambertian radiator.

Different Forms

Planck's law can be encountered in several forms depending on the conventions and preferences of different scientific fields. The various forms of the law for spectral radiance are summarized in the table below. Forms on the left are most often encountered in experimental fields, while those on the right are most often encountered in theoretical fields.

Planck's law expressed in terms of different spectral variables			
with h		***with ?***	
variable	***distribution***	***variable***	***distribution***
Frequency ν	$B_\nu(\nu,T)=\frac{2h\nu^3}{c^2}\frac{1}{e^{h\nu/(k_BT)}-1}$	Angular frequency ω	$B_\omega(\omega,T)=\frac{\hbar\omega^3}{4\pi^3c^2}\frac{1}{e^{\hbar\omega/(k_BT)}-1}$
Wavelength λ	$B_\lambda(\lambda,T)=\frac{2hc^2}{\lambda^5}\frac{1}{e^{hc/(\lambda k_BT)}-1}$	Angular wavelength y	$B_y(y,T)=\frac{\hbar c^2}{4\pi^3y^5}\frac{1}{e^{\hbar c/(yk_BT)}-1}$
Wavenumber $\tilde{\nu}$	$B_{\tilde{\nu}}(\tilde{\nu},T)=2hc^2\tilde{\nu}^3\frac{1}{e^{hc\tilde{\nu}/(k_BT)}-1}$	Angular wavenumber k	$B_k(k,T)=\frac{\hbar c^2k^3}{4\pi^3}\frac{1}{e^{\hbar ck/(k_BT)}-1}$

These distributions represent the spectral radiance of blackbodies—the power emitted from the emitting surface, per unit projected area of emitting surface, per unit solid angle, per spectral unit (frequency, wavelength, wavenumber or their angular equivalents). Since the radiance is isotropic (i.e. independent of direction), the power emitted at an angle to the normal is proportional to the projected area, and therefore to the cosine of that angle as per Lambert's cosine law, and is unpolarized.

Correspondence between Spectral Variable Forms

Different spectral variables require different corresponding forms of expression of the law. In general, one may not convert between the various forms of Planck's law simply by substituting one variable for another, because this would not take into account that the different forms have different units. Wavelength and frequency units are reciprocal.

Corresponding forms of expression are related because they express one and the same physical fact: for a particular physical spectral increment, a corresponding particular physical energy increment is radiated.

This is so whether it is expressed in terms of an increment of frequency, dν, or, correspondingly, of wavelength, dλ. Introduction of a minus sign can indicate that an increment of frequency corresponds with decrement of wavelength. For the above corresponding forms of expression of the spectral radiance, one may use an obvious expansion of notation, temporarily for the present calculation only. Then, for a particular spectral increment, the particular physical energy increment may be written

$$B_\lambda(\lambda, T)\,d\lambda = -B_\nu(\nu(\lambda), T)\,d\nu, \text{ which leads to}$$

$$B_\lambda(\lambda, T) = -\frac{d\nu}{d\lambda}B_\nu(\nu(\lambda), T).$$

Also, $\nu(\lambda) = c/\lambda$, so that $d\nu/d\lambda = -c/\lambda^2$. Substitution gives the correspondence between the frequency and wavelength forms, with their different units.

It follows that the location of the peak of the distribution for Planck's law depends on the choice of spectral variable.

Spectral Energy Density Form

Planck's law can also be written in terms of the spectral energy density (u) by multiplying B by $4\pi/c$:

$$u_i(T) = \frac{4\pi}{c} B_i(T).$$

These distributions have units of energy per volume per spectral unit.

Derivation

Consider a cube of side L with conducting walls filled with electromagnetic radiation in thermal equilibrium at temperature T. If there is a small hole in one of the walls, the radiation emitted from the hole will be characteristic of a perfect black body. We will first calculate the spectral energy density within the cavity and then determine the spectral radiance of the emitted radiation.

At the walls of the cube, the parallel component of the electric field and the orthogonal component of the magnetic field must vanish. Analogous to the wave function of a particle in a box, one finds that the fields are superpositions of periodic functions. The three wavelengths λ_1, λ_2, and λ_3, in the three directions orthogonal to the walls can be:

$$\lambda_i = \frac{2L}{n_i},$$

where the n_i are integers. For each set of integers n_i there are two linear independent solutions (modes). According to quantum theory, the energy levels of a mode are given by:

$$E_{n_1,n_2,n_3}(r) = \left(r + \frac{1}{2}\right)\frac{hc}{2L}\sqrt{n_1^2 + n_2^2 + n_3^2}. \qquad (1)$$

The quantum number r can be interpreted as the number of photons in the mode. The two modes for each set of n_i correspond to the two polarization states of the photon which has a spin of 1. Note that for $r = 0$ the energy of the mode is not zero. This vacuum energy of the electromagnetic field is responsible for the Casimir effect. In the following we will calculate the internal energy of the box at absolute temperature T.

According to statistical mechanics, the probability distribution over the energy levels of a particular mode is given by:

$$P_r = \frac{\exp(-\beta E(r))}{Z(\beta)}.$$

Here

$$\beta \stackrel{\text{def}}{=} 1/(k_B T).$$

The denominator $Z(\beta)$, is the partition function of a single mode and makes P_r properly normalized:

$$Z(\beta) = \sum_{r=0}^{\infty} e^{-\beta E(r)} = \frac{e^{-\beta\varepsilon/2}}{1 - e^{-\beta\varepsilon}}.$$

Here we have implicitly defined

$$\varepsilon \stackrel{\text{def}}{=} \frac{hc}{2L}\sqrt{n_1^2 + n_2^2 + n_3^2},$$

which is the energy of a single photon. As explained here, the average energy in a mode can be expressed in terms of the partition function:

$$\langle E \rangle = -\frac{d\log(Z)}{d\beta} = \frac{\varepsilon}{2} + \frac{\varepsilon}{e^{\beta\varepsilon} - 1}.$$

This formula, apart from the first vacuum energy term, is a special case of the general formula for particles obeying Bose–Einstein statistics. Since there is no restriction on the total number of photons, the chemical potential is zero.

If we measure the energy relative to the ground state, the total energy in the box follows by summing $\langle E \rangle - \frac{\varepsilon}{2}$ over all allowed single photon states. This can be done exactly in the thermodynamic limit as L approaches infinity. In this limit, ε becomes continuous and we can then integrate $\langle E \rangle - \frac{\varepsilon}{2}$ over this parametre. To calculate the energy in the box in this way, we need to evaluate how many photon states there are in a given energy range. If we write the total number of single photon states with energies between ε and $\varepsilon + d\varepsilon$ as $g(\varepsilon)d\varepsilon$, where $g(\varepsilon)$ is the density of states (which we'll evaluate in a moment), then we can write:

$$U = \int_0^{\infty} \frac{\varepsilon}{e^{\beta\varepsilon} - 1} g(\varepsilon)d\varepsilon. \qquad (2)$$

To calculate the density of states we rewrite equation (1) as follows:

$$\varepsilon \stackrel{\text{def}}{=} \frac{hc}{2L}n,$$

where n is the norm of the vector n = (n_1, n_2, n_3):

$$n = \sqrt{n_1^2 + n_2^2 + n_3^2}.$$

For every vector n with integer components larger than or equal to zero, there are two photon states. This means that the number of photon states in a certain region of n-space is twice the volume of that region. An energy range of $d\varepsilon$ corresponds to shell of thickness $dn = (2L/hc)d\varepsilon$ in n-space. Because the components of n have to be positive, this shell spans an octant of a sphere.

The number of photon states $g(\varepsilon)d\varepsilon$, in an energy range $d\varepsilon$, is thus given by:

$$g(\varepsilon)d\varepsilon = 2\frac{1}{8}4\pi n^2\, dn = \frac{8\pi L^3}{h^3c^3}\varepsilon^2\, d\varepsilon.$$

Inserting this in Eq. (2) gives:

$$U = L^3\frac{8\pi}{h^3c^3}\int_0^\infty \frac{\varepsilon^3}{e^{\beta\varepsilon}-1}d\varepsilon. \tag{3}$$

From this equation one can derive the spectral energy density as a function of frequency $u_\nu(T)$ and as a function of wavelength $u_\lambda(T)$:

$$\frac{U}{L^3} = \int_0^\infty u_\nu(T)d\nu,$$

where:

$$u_\nu(T) = \frac{8\pi h\nu^3}{c^3}\frac{1}{e^{h\nu/k_\mathrm{B}T}-1}.$$

And:

$$\frac{U}{L^3} = \int_0^\infty u_\lambda(T)d\lambda,$$

where

$$u_\lambda(T) = \frac{8\pi hc}{\lambda^5}\frac{1}{e^{hc/\lambda k_\mathrm{B}T}-1}.$$

This is also a spectral energy density function with units of energy per unit wavelength per unit volume. Integrals of this type for Bose and Fermi gases can be expressed in terms of polylogarithms. In this case, however, it is possible to calculate the integral in closed form using only elementary functions. Substituting

$$\varepsilon = k_\mathrm{B}Tx,$$

in Eq. (3), makes the integration variable dimensionless giving:

$$u(T) = \frac{8\pi(k_\mathrm{B}T)^4}{(hc)^3}J,$$

where J is a Bose–Einstein integral given by:

$$J = \int_0^\infty \frac{x^3}{e^x-1}dx = \frac{\pi^4}{15}.$$

The total electromagnetic energy inside the box is thus given by:

$$\frac{U}{V} = \frac{8\pi^5 (k_B T)^4}{15(hc)^3},$$

where $V = L^3$ is the volume of the box.

The combination hc/k_B has the value 14 387.770 μm ·K.

This is not the Stefan–Boltzmann law (which provides the total energy *radiated* by a black body per unit surface area per unit time), but it can be written more compactly using the Stefan–Boltzmann constant σ, giving

$$\frac{U}{V} = \frac{4\sigma T^4}{c}.$$

The constant $4\sigma/c$ is sometimes called the radiation constant.

Since the radiation is the same in all directions, and propagates at the speed of light (c), the spectral radiance of radiation exiting the small hole is

$$B_\nu(T) = \frac{u_\nu(T)c}{4\pi},$$

which yields

$$B_\nu(T) = \frac{2h\nu^3}{c^2}\frac{1}{e^{h\nu/k_B T} - 1}.$$

It can be converted to an expression for $B_\lambda(T)$ in wavelength units by substituting υ by c/λ and evaluating

$$B_\lambda(T) = B_\nu(T)\left|\frac{d\nu}{d\lambda}\right|.$$

Note that dimensional analysis shows that the unit of steradians, shown in the denominator of left hand side of the equation above, is generated in and carried through the derivation but does not appear in any of the dimensions for any element on the left-hand-side of the equation. This derivation is based on Brehm & Mullin 1989.

Physics

Outline

Planck's law describes the unique and characteristic spectral distribution for electromagnetic radiation in thermodynamic equilibrium, when there is no net flow of matter or energy. Its physics is most easily understood by considering the radiation in a cavity with rigid opaque walls.

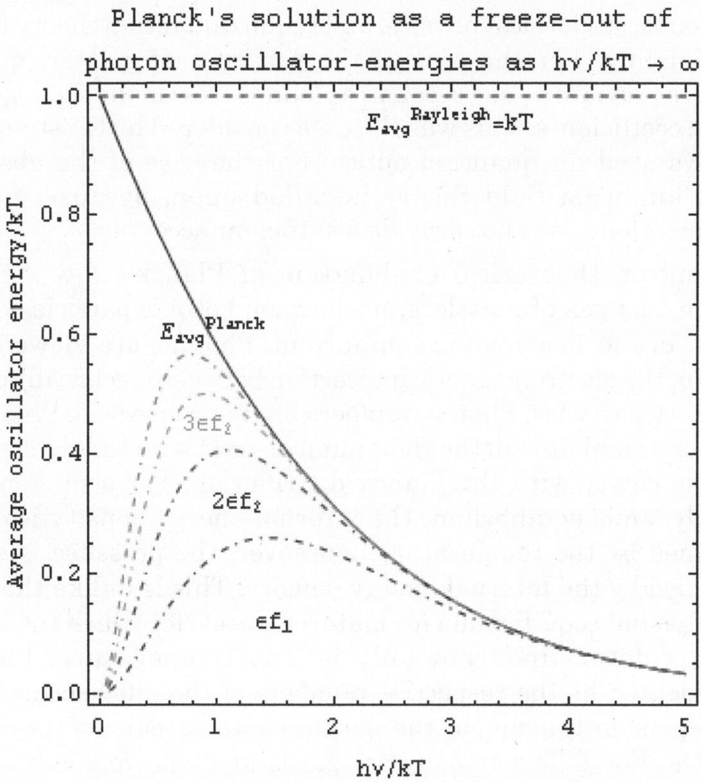

Figure: *Freeze-out of high-energy oscillators.*

Motion of the walls can affect the radiation. If the walls are not opaque, then the thermodynamic equilibrium is not isolated. It is of interest to explain how the thermodynamic equilibrium is attained. There are two main cases:

(a) when the approach to thermodynamic equilibrium is in the presence of matter, when the walls of the cavity are imperfectly reflective for every wavelength or when the walls are perfectly reflective while the cavity contains a small black body (this was the main case considered by Planck); or

(b) when the approach to equilibrium is in the absence of matter, when the walls are perfectly reflective for all wavelengths and the cavity contains no matter. For matter not enclosed in such a cavity, thermal radiation can be approximately explained by appropriate use of Planck's law.

Classical physics provides an account of some aspects of the Planck distribution, such as the Stefan–Boltzmann law, and the Wien displacement law. Other aspects of the Planck distribution cannot be accounted for in classical physics, and require quantum theory for their understanding. For the case of the presence of matter, quantum mechanics provides a good account, as found below in the section headed Einstein coefficients. This was the case considered by Einstein, and is nowadays used for quantum optics. For the case of the absence of matter, quantum field theory is called upon, because quantum mechanics alone does not provide a sufficient account.

Quantum theoretical explanation of Planck's law views the radiation as a gas of massless, uncharged, bosonic particles, namely photons, in thermodynamic equilibrium. Photons are viewed as the carriers of the electromagnetic interaction between electrically charged elementary particles. Photon numbers are not conserved. Photons are created or annihilated in the right numbers and with the right energies to fill the cavity with the Planck distribution. For a photon gas in thermodynamic equilibrium, the internal energy density is entirely determined by the temperature; moreover, the pressure is entirely determined by the internal energy density. This is unlike the case of thermodynamic equilibrium for material gases, for which the internal energy is determined not only by the temperature, but also, independently, by the respective numbers of the different molecules, and independently again, by the specific characteristics of the different molecules. For different material gases at given temperature, the pressure and internal energy density can vary independently, because different molecules can carry independently different excitation energies.

Planck's law arises as a limit of the Bose–Einstein distribution, the energy distribution describing non-interactive bosons in thermodynamic equilibrium. In the case of massless bosons such as photons and gluons, the chemical potential is zero and the Bose-Einstein distribution reduces to the Planck distribution. There is another fundamental equilibrium energy distribution: the Fermi–Dirac distribution, which describes fermions, such as electrons, in thermal equilibrium. The two distributions differ because multiple bosons can occupy the same quantum state, while multiple fermions cannot. At low densities, the number of available quantum states per particle is large, and this difference becomes irrelevant. In the low density limit, the Bose-Einstein and the Fermi-Dirac distribution each reduce to the Maxwell–Boltzmann distribution.

Kirchhoff's Law of Thermal Radiation

Kirchhoff's law of thermal radiation is a succinct and brief account of a complicated physical situation. The following is an introductory sketch of that situation, and is very far from being a rigorous physical argument. The purpose here is only to summarize the main physical factors in the situation, and the main conclusions.

Spectral Dependence of Thermal Radiation

There is a difference between conductive heat transfer and radiative heat transfer. Radiative heat transfer can be filtered to pass only a definite band of radiative frequencies.

It is generally known that the hotter a body becomes the more heat it radiates at every frequency.

In a cavity in an opaque body with rigid walls that are not perfectly reflective at any frequency, in thermodynamic equilibrium, there is only one temperature, and it must be shared in common by the radiation of every frequency.

One may imagine two such cavities, each in its own isolated radiative and thermodynamic equilibrium. One may imagine an optical device that allows radiative heat transfer between the two cavities, filtered to pass only a definite band of radiative frequencies. If the values of the spectral radiances of the radiations in the cavities differ in that frequency band, heat may be expected to pass from the hotter to the colder. One might propose to use such a filtered transfer of heat in such a band to drive a heat engine. If the two bodies are at the same temperature, the second law of thermodynamics does not allow the heat engine to work. It may be inferred that for a temperature common to the two bodies, the values of the spectral radiances in the pass-band must also be common. This must hold for every frequency band. This became clear to Balfour Stewart and later to Kirchhoff. Balfour Stewart found experimentally that of all surfaces, one of lamp-black emitted the greatest amount of thermal radiation for every quality of radiation, judged by various filters.

Thinking theoretically, Kirchhoff went a little further, and pointed out that this implied that the spectral radiance, as a function of radiative frequency, of any such cavity in thermodynamic equilibrium must be a unique universal function of temperature. He postulated an ideal black body that interfaced with its surrounds in just such a way as to absorb all the radiation that falls on it. By the Helmholtz reciprocity principle, radiation from the interior of such a body would pass unimpeded, directly to its surrounds without reflection at the interface. In thermodynamic equilibrium, the thermal radiation emitted from such a body would have that unique universal spectral radiance as a function of temperature. This insight is the root of Kirchhoff's law of thermal radiation.

Relation between Absorptivity and Emissivity

One may imagine a small homogeneous spherical material body labelled *X* at a temperature T_X, lying in a radiation field within a large cavity with walls of material labelled *Y* at a temperature T_Y. The body *X* emits its own thermal radiation.

At a particular frequency ν, the radiation emitted from a particular cross-section through the centre of *X* in one sense in a direction normal to that cross-section may be denoted $I_{\nu,X}(T_X)$, characteristically for the material of *X*. At that frequency ν, the radiative power from the walls into that cross-section in the opposite sense in that direction may be denoted $I_{\nu,Y}(T_Y)$, for the wall temperature T_Y. For the material of *X*, defining the absorptivity $\alpha_{\nu,X,Y}(T_X, T_Y)$ as the fraction of that incident radiation absorbed by *X*, that incident energy is absorbed at a rate $\alpha_{\nu,X,Y}(T_X, T_Y)\, I_{\nu,Y}(T_Y)$.

The rate $q(\nu, T_X, T_Y)$ of accumulation of energy in one sense into the cross-section of the body can then be expressed

$$q(\nu,T_X,T_Y)=\alpha_{\nu,X,Y}(T_X,T_Y)I_{\nu,Y}(T_Y)-I_{\nu,X}(T_X).$$

Kirchhoff's seminal insight, mentioned just above, was that, at thermodynamic equilibrium at temperature *T*, there exists a unique universal radiative distribution, nowadays denoted $B_\nu(T)$, that is independent of the chemical characteristics of the materials *X* and *Y*, that leads to a very valuable understanding of the radiative exchange equilibrium of any body at all, as follows.

When there is thermodynamic equilibrium at temperature *T*, the cavity radiation from the walls has that unique universal value, so that $I_{\nu,Y}(T_Y) = B_\nu(T)$. Further, one may define the emissivity $\varepsilon_{\nu,X}(T_X)$ of the material of the body *X* just so that at thermodynamic equilibrium at temperature $T_X = T$, one has $I_{\nu,X}(T_X) = I_{\nu,X}(T) = \varepsilon_{\nu,X}(T)\, B_\nu(T)$.

When thermal equilibrium prevails at temperature $T = T_X = T_Y$, the rate of accumulation of energy vanishes so that $q(\nu,T_X,T_Y) = 0$. It follows that:

$$\text{in thermodynamic equilibrium, when } T=T_X=T_Y,$$
$$\text{it is true that } 0=\alpha_{\nu,X,Y}(T,T)B_\nu(T)-\epsilon_{\nu,X}(T)B_\nu(T).$$

Kirchhoff pointed out that it follows that

$$\text{in thermodynamic equilibrium, when } T=T_X=T_Y$$
$$\text{, it is true that } \alpha_{\nu,X,Y}(T,T)=\epsilon_{\nu,X}(T).$$

Introducing the special notation $\alpha_{\nu,X}(T)$ for the absorptivity of material *X* at thermodynamic equilibrium at temperature *T* (justified by a discovery of Einstein, as indicated below), one further has the equality

$$\alpha_{\nu,X}(T) = \epsilon_{\nu,X}(T) \text{ at thermodynamic equilibrium.}$$

The equality of absorptivity and emissivity here demonstrated is specific for thermodynamic equilibrium at temperature T and is in general not to be expected to hold when conditions of thermodynamic equilibrium do not hold. The emissivity and absorptivity are each separately properties of the molecules of the material but they depend differently upon the distributions of states of molecular excitation on the occasion, because of a phenomenon known as "stimulated emission", that was discovered by Einstein. On occasions when the material is in thermodynamic equilibrium or in a state known as local thermodynamic equilibrium, the emissivity and absorptivity become equal. Very strong incident radiation or other factors can disrupt thermodynamic equilibrium or local thermodynamic equilibrium. Local thermodynamic equilibrium in a gas means that molecular collisions far outweigh light emission and absorption in determining the distributions of states of molecular excitation.

Kirchhoff pointed out that he did not know the precise character of $B_\nu(T)$, but he thought it important that it should be found out. Four decades after Kirchhoff's insight of the general principles of its existence and character, Planck's contribution was to determine the precise mathematical expression of that equilibrium distribution $B_\nu(T)$.

Black Body

In physics, one considers an ideal black body, here labelled B, defined as one that completely absorbs all of the electromagnetic radiation falling upon it at every frequency ν (hence the term "black"). According to Kirchhoff's law of thermal radiation, this entails that, for every frequency ν, at thermodynamic equilibrium at temperature T, one has $\alpha\nu_{,B}(T) = \varepsilon\nu_{,B}(T) = 1$, so that the thermal radiation from a black body is always equal to the full amount specified by Planck's law. No physical body can emit thermal radiation that exceeds that of a black body, since if it were in equilibrium with a radiation field, it would be emitting more energy than was incident upon it.

Though perfectly black materials do not exist, in practice a black surface can be accurately approximated. As to its material interior, a body of condensed matter, liquid, solid, or plasma, with a definite interface with its surroundings, is completely black to radiation if it is completely opaque. That means that it absorbs all of the radiation that penetrates the interface of the body with its surroundings, and enters the body. This is not too difficult to achieve in practice. On the other hand, a perfectly black interface is not found in nature. A perfectly black interface reflects no radiation, but transmits all that falls on it, from either side.

The best practical way to make an effectively black interface is to simulate an 'interface' by a small hole in the wall of a large cavity in a completely opaque rigid body of material that does not reflect perfectly at any frequency, with its walls at a controlled temperature. Beyond these requirements, the component material of the walls is unrestricted. Radiation entering the hole has almost no possibility of escaping the cavity without being absorbed by multiple impacts with its walls.

Lambert's Cosine Law

As explained by Planck, a radiating body has an interior consisting of matter, and an interface with its contiguous neighbouring material medium, which is usually the medium from within which the radiation from the surface of the body is observed. The interface is not composed of physical matter but is a theoretical conception, a mathematical two-dimensional surface, a joint property of the two contiguous media, strictly speaking belonging to neither separately.

Such an interface can neither absorb nor emit, because it is not composed of physical matter; but it is the site of reflection and transmission of radiation, because it is a surface of discontinuity of optical properties. The reflection and transmission of radiation at the interface obey the Stokes–Helmholtz reciprocity principle.

At any point in the interior of a black body located inside a cavity in thermodynamic equilibrium at temperature T the radiation is homogeneous, isotropic and unpolarized. A black body absorbs all and reflects none of the electromagnetic radiation incident upon it. According to the Helmholtz reciprocity principle, radiation from the interior of a black body is not reflected at its surface, but is fully transmitted to its exterior.

Because of the isotropy of the radiation in the body's interior, the spectral radiance of radiation transmitted from its interior to its exterior through its surface is independent of direction.

This is expressed by saying that radiation from the surface of a black body in thermodynamic equilibrium obeys Lambert's cosine law. This means that the spectral flux $d\vartheta(dA, \theta, d\Omega, d\nu)$ from a given infinitesimal element of area dA of the actual emitting surface of the black body, detected from a given direction that makes an angle θ with the normal to the actual emitting surface at dA, into an element of solid angle of detection dΩ centred on the direction indicated by θ, in an element of frequency bandwidth dν, can be represented as

$$\frac{\mathrm{d}\Phi(\mathrm{d}A,\theta,\mathrm{d}\Omega,\mathrm{d}\nu)}{\mathrm{d}\Omega}=L^0(\mathrm{d}A,\mathrm{d}\nu)\mathrm{d}A\mathrm{d}\nu\cos\theta$$

where $L^0(\mathrm{d}A, \mathrm{d}\nu)$ denotes the flux, per unit area per unit frequency per unit solid angle, that area dA would show if it were measured in its normal direction $\theta = 0$.

The factor cos θ is present because the area to which the spectral radiance refers directly is the projection, of the actual emitting surface area, onto a plane perpendicular to the direction indicated by θ . This is the reason for the name *cosine law*.

Taking into account the independence of direction of the spectral radiance of radiation from the surface of a black body in thermodynamic equilibrium, one has $L^0(\mathrm{d}A, \mathrm{d}\nu) = B_\nu(T)$ and so

$$\frac{\mathrm{d}\Phi(\mathrm{d}A,\theta,\mathrm{d}\Omega,\mathrm{d}\nu)}{\mathrm{d}\Omega}=B_\nu(T)\mathrm{d}A\mathrm{d}\nu\cos\theta.$$

Thus Lambert's cosine law expresses the independence of direction of the spectral radiance $B_\nu(T)$ of the surface of a black body in thermodynamic equilibrium.

Stefan–Boltzmann Law

The total power emitted per unit area at the surface of a black body (P) may be found by integrating the black body spectral flux found from Lambert's law over all frequencies, and over the solid angles corresponding to a hemisphere (h) above the surface.

$$P=\int_0^\infty d\nu\int_h d\Omega B_\nu\cos(\theta)$$

The infinitesimal solid angle can be expressed in spherical polar coordinates:

$$d\Omega=\sin(\theta)d\theta d\phi.$$

So that:

$$P=\int_0^\infty d\nu\int_0^{\pi/2}d\theta\int_0^{2\pi}d\phi B_\nu(T)\cos(\theta)\sin(\theta)=\sigma T^4$$

where

$$\sigma=\frac{2k_B^4\pi^5}{15c^2h^3}\approx 5.670400\times10^{-8}\,\mathrm{Js^{-1}m^{-2}K^{-4}}$$

is known as the Stefan–Boltzmann constant.

Radiative Transfer

The equation of radiative transfer describes the way in which radiation is affected as it travels through a material medium. For the special case in which the material medium is in thermodynamic equilibrium in the neighbourhood of a point in the medium, Planck's law is of special importance.

For simplicity, we can consider the linear steady state, without scattering. The equation of radiative transfer states that for a beam of light going through a small distance *ds*, energy is conserved: The change in the (spectral) radiance of that beam (I_ν) is equal to the amount removed by the material medium plus the amount gained from the material medium.

If the radiation field is in equilibrium with the material medium, these two contributions will be equal. The material medium will have a certain emission coefficient and absorption coefficient.

The absorption coefficient α is the fractional change in the intensity of the light beam as it travels the distance *ds*, and has units of 1/length. It is composed of two parts, the decrease due to absorption and the increase due to stimulated emission. Stimulated emission is emission by the material body which is caused by and is proportional to the incoming radiation.

It is included in the absorption term because, like absorption, it is proportional to the intensity of the incoming radiation. Since the amount of absorption will generally vary linearly as the density ρ of the material, we may define a "mass absorption coefficient" $\kappa_\nu = \alpha / \rho$ which is a property of the material itself. The change in intensity of a light beam due to absorption as it traverses a small distance *ds* will then be $dI_\nu = -\kappa_\nu \rho I_\nu ds$

The "mass emission coefficient" j_ν is equal to the radiance per unit volume of a small volume element divided by its mass (since, as for the mass absorption coefficient, the emission is proportional to the emitting mass) and has units of power/solid angle/frequency/density. Like the mass absorption coefficient, it too is a property of the material itself. The change in a light beam as it traverses a small distance *ds* will then be $dI_\nu = j_\nu \rho ds$

The equation of radiative transfer will then be the sum of these two contributions:

$$\frac{dI_\nu}{ds} = j_\nu \rho - \kappa_\nu \rho I_\nu .$$

If the radiation field is in equilibrium with the material medium, then the radiation will be homogeneous (independent of position) so that $dI_\nu = 0$ and:

$$\kappa_\nu B_\nu = j_\nu$$

which is another statement of Kirchhoff's law, relating two material properties of the medium, and which yields the radiative transfer equation at a point around which the medium is in thermodynamic equilibrium:

$$\frac{dI_\nu}{ds} = \kappa_\nu \rho (B_\nu - I_\nu).$$

Einstein Coefficients

The principle of detailed balance states that, at thermodynamic equilibrium, each elementary process is equilibrated by its reverse process.

In 1916, Albert Einstein applied this principle on an atomic level to the case of an atom radiating and absorbing radiation due to transitions between two particular energy levels, giving a deeper insight into the equation of radiative transfer and Kirchhoff's law for this type of radiation. If level 1 is the lower energy level with energy E_1, and level 2 is the upper energy level with energy E_2, then the frequency $\nu$of the radiation radiated or absorbed will be determined by Bohr's frequency condition: $E_2 - E_1 = h\nu$.

If n_1 and n_2 are the number densities of the atom in states 1 and 2 respectively, then the rate of change of these densities in time will be due to three processes:

$\left(\frac{dn_1}{dt}\right)_{\text{spon}} = A_{21} n_2$	Spontaneous emission
$\left(\frac{dn_1}{dt}\right)_{\text{stim}} = B_{21} n_2 I_\nu(T)$	Stimulated emission
$\left(\frac{dn_2}{dt}\right)_{\text{abs}} = B_{12} n_1 I_\nu(T)$	Photo-absorption

where $I_\nu(T)$ is the spectral radiance of the radiation field. The three parametres A_{21}, B_{21} and B_{12}, known as the Einstein coefficients, are associated with the photon frequency (ν) produced by the transition between two energy levels (states). As a result, each line in a spectra has it own set of associated coefficients. When the atoms and the radiation field are in equilibrium, the radiance will be given by Planck's law and, by the principle of detailed balance, the sum of these rates must be zero:

$$0 = A_{21}n_2 + B_{21}n_2B_\nu(T) - B_{12}n_1B_\nu(T)$$

Since the atoms are also in equilibrium, the populations of the two levels are related by the Boltzmann factor:

$$\frac{n_2}{n_1} = \frac{g_2}{g_1}e^{-h\nu/k_BT}$$

where g_1 and g_2 are the multiplicities of the respective energy levels. Combining the above two equations with the requirement that they be valid at any temperature yields two relationships between the Einstein coefficients:

$$\frac{A_{21}}{B_{21}} = \frac{2h\nu^3}{c^2}$$

$$\frac{B_{21}}{B_{12}} = \frac{g_1}{g_2}$$

so that knowledge of one coefficient will yield the other two. For the case of isotropic absorption and emission, the emission coefficient (j_ν) and absorption coefficient (κ_ν) defined in the radiative transfer section above, can be expressed in terms of the Einstein coefficients. The relationships between the Einstein coefficients will yield the expression of Kirchhoff's law expressed in the *Radiative transfer* section above, namely that

$$j_\nu = \kappa_\nu B_\nu.$$

These coefficients apply to both atoms and molecules.

Properties

Peaks

The distributions $B_\nu, B_\omega, B_{\tilde{\nu}}$ and B_k peak at a photon energy of

$$E = \left[3 + W\left(\frac{-3}{e^3}\right)\right]k_BT \approx 2.821\,k_BT,$$

where W is the Lambert W function. The distributions B_λ and B_y however, peak at a different energy

$$E = \left[5 + W\left(\frac{-5}{e^5}\right)\right]k_BT \approx 4.965\,k_BT,$$

The reason for this is that, as mentioned above, one cannot go from (for example) B_ν to B_λ simply by substituting ν by λ. In addition, one must also multiply the result of the substitution by $\left|\frac{d\nu}{d\lambda}\right| = c/\lambda^2$. This $1/\lambda^2$ factor shifts the peak of the distribution to higher energies.

Dividing *hc* by this energy expression gives the wavelength of the peak. For this one can use $hc/k_B = 14\,387.770$ μm·K.

The spectral radiance at these peaks is given by:

$$B_{\nu,\max}(T) = \frac{2k_B^3T^3(3+W(-3\exp(-3)))^3}{h^2c^2}\frac{1}{e^{3+W(-3\exp(-3))}-1} \approx 1.896\times10^{-19}\ \text{watt m}^{-2}\text{Hz}^{-1}\text{kelvin}^{-3}T^3$$

$$B_{\lambda,\max}(T) = \frac{2k_B^5T^5(5+W(-5\exp(-5)))^5}{h^4c^3}\frac{1}{e^{5+W(-5\exp(-5))}-1} \approx 4.09567\times10^{-6}\ \text{watt m}^{-2}\text{m}^{-1}\text{kelvin}^{-5}T^5$$

Approximations

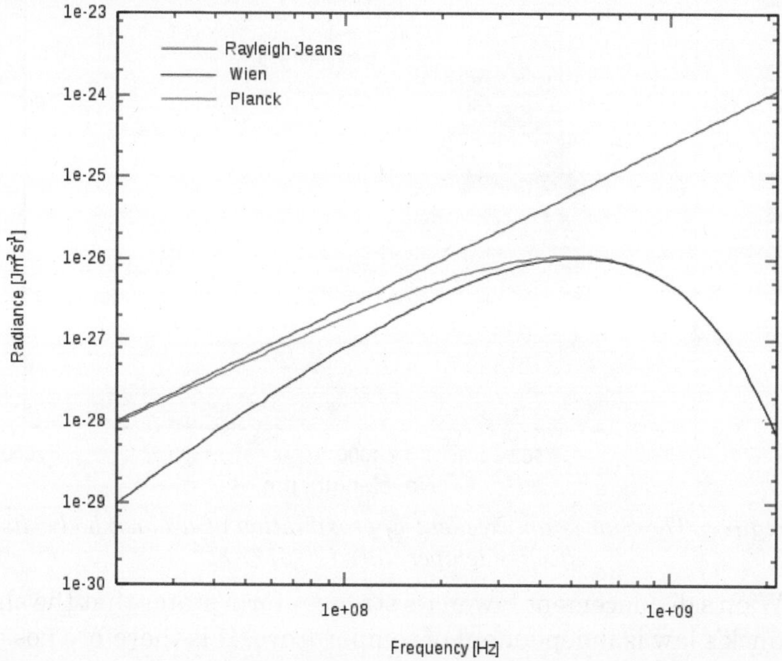

Figure: *Log-log plots of radiance vs. frequency for Planck's law (green), compared with the Rayleigh–Jeans law (red) and the Wien approximation (blue) for a black body at 8 mK temperature.*

In the limit of low frequencies (i.e. long wavelengths), Planck's law becomes the Rayleigh–Jeans law

$$B_\nu(T) \approx \frac{2\nu^2}{c^2}k_BT \quad \text{or} \quad B_\lambda(T) \approx \frac{2c}{\lambda^4}k_BT.$$

The radiance increases as the square of the frequency, illustrating the ultraviolet catastrophe.

In the limit of high frequencies (i.e. small wavelengths) Planck's law tends to the Wien approximation:

$$B_\nu(T) \approx \frac{2h\nu^3}{c^2} e^{-\frac{h\nu}{k_B T}} \quad \text{or} \quad B_\lambda(T) \approx \frac{2hc^2}{\lambda^5} e^{-\frac{hc}{\lambda k_B T}}.$$

Both approximations were known to Planck before he developed his law. He was led by these two approximations to develop a law which incorporated both limits, which ultimately became Planck's law.

Percentiles

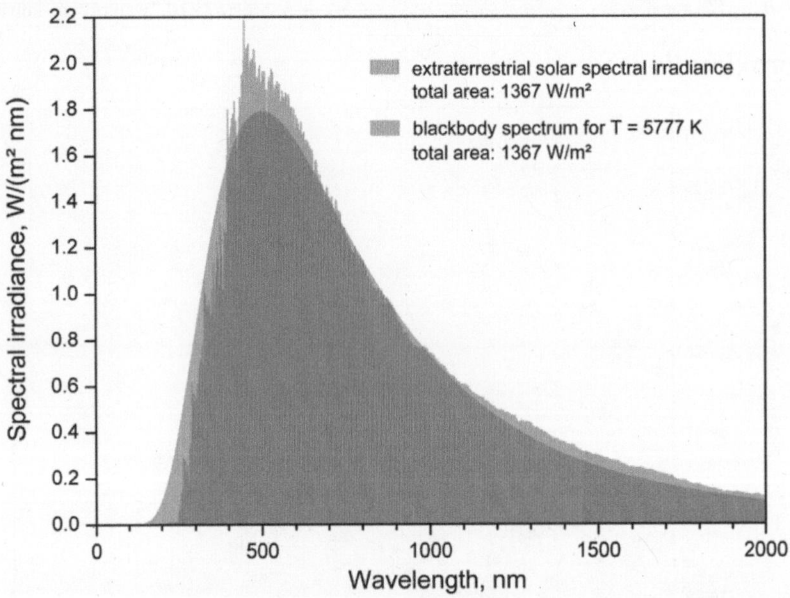

Figure: *The Sun is an excellent approximation of a black body. Its effective temperature is ~5777 K.*

Wien's displacement law in its stronger form states that the shape of Planck's law is independent of temperature. It is therefore possible to list the percentile points of the total radiation as well as the peaks for wavelength and frequency, in a form which gives the wavelength λ when divided by temperature T.

The second row of the following table lists the corresponding values of λT, that is, those values of x for which the wavelength λ is x/T micrometres at the radiance percentile point given by the corresponding entry in the first row.

Percentile	0.01%	0.1%	1%	10%	20%	25.0%	30%	40%	41.8%	50%	60%	64.6%	70%	80%	90%	99%	99.9%	99.99%
λT (μm · K)	910	1110	1448	2195	2676	2898	3119	3582	3670	4107	4745	5099	5590	6864	9376	22884	51613	113374

That is, 0.01% of the radiation is at a wavelength below $910/T$ μm, 20% below $2676/T$ μm, etc. The wavelength and frequency peaks are in bold and occur at 25.0% and 64.6% respectively.

The 41.8% point is the wavelength-frequency-neutral peak. These are the points at which the respective Planck-law functions $1/\lambda^5$, ν^3, and ν^2/λ^2 divided by $\exp(h\nu/k_BT) - 1$ attain their maxima. Also note the much smaller gap in ratio of wavelengths between 0.1% and 0.01% (1110 is 22% more than 910) than between 99.9% and 99.99% (113374 is 120% more than 51613), reflecting the exponential decay of energy at short wavelengths (left end) and polynomial decay at long.

Which peak to use depends on the application. The conventional choice is the wavelength peak at 25.0% given by Wien's displacement law in its weak form. For some purposes the median or 50% point dividing the total radiation into two halves may be more suitable.

The latter is closer to the frequency peak than to the wavelength peak because the radiance drops exponentially at short wavelengths and only polynomially at long. The neutral peak occurs at a shorter wavelength than the median for the same reason.

For the Sun, T is 5778 K, allowing the percentile points of the Sun's radiation, in manometres, to be tabulated as follows when modelled as a black body radiator, to which the Sun is a fair approximation. For comparison a planet modelled as a black body radiating at a nominal 288 K (15 °C) as a representative value of the Earth's highly variable temperature has wavelengths more than twenty times that of the Sun, tabulated in the third row in micrometres (thousands of manometres).

Percentile	0.01%	0.1%	1%	10%	20%	25.0%	30%	40%	41.8%	50%	60%	64.6%	70%	80%	90%	99%	99.9%	99.99%
Sun λ (nm)	157	192	251	380	463	502	540	620	635	711	821	882	967	1188	1623	3961	8933	19620
288 K planet λ (μm)	3.16	3.85	5.03	7.62	9.29	10.1	10.8	12.4	12.7	14.3	16.5	17.7	19.4	23.8	32.6	79.5	179	394

That is, only 1% of the Sun's radiation is at wavelengths shorter than 251 nm, and only 1% at longer than 3961 nm. Expressed in micrometres this puts 98% of the Sun's radiation in the range from 0.251 to 3.961 μm. The corresponding 98% of energy radiated from a 288 K planet is from 5.03 to 79.5 μm, well above the range of solar radiation (or below if expressed in terms of frequencies $\nu = c/\lambda$ instead of wavelengths λ).

A consequence of this more-than-order-of-magnitude difference in wavelength between solar and planetary radiation is that filters designed to pass one and block the other are easy to construct. For example windows fabricated of ordinary glass or transparent plastic pass at least 80% of the incoming 5778 K solar radiation, which is below 1.2 μm in wavelength, while blocking over 99% of the outgoing 288 K thermal radiation from 5 μm upwards, wavelengths at which most kinds of glass and plastic of construction-grade thickness are effectively opaque.

The Sun's radiation is that arriving at the top of the atmosphere (TOA). As can be read from the table, radiation below 400 nm, or ultraviolet, is about 12%, while that above 700 nm, or infrared, starts at about the 49% point and so accounts for 51% of the total. Hence only 37% of the TOA insolation is visible to the human eye. The atmosphere shifts these percentages substantially in favour of visible light as it absorbs most of the ultraviolet and significant amounts of infrared.

History

Forerunners

Balfour Stewart: In 1858, Balfour Stewart described his experiments on the thermal radiative emissive and absorptive powers of polished plates of various substances, compared with the powers of lamp-black surfaces, at the same temperature. Stewart chose lamp-black surfaces as his reference because of various previous experimental findings, especially those of Pierre Prevost and of John Leslie. He wrote "Lamp-black, which absorbs all the rays that fall upon it, and therefore possesses the greatest possible absorbing power, will possess also the greatest possible radiating power."

Stewart measured radiated power with a thermo-pile and sensitive galvanometre read with a microscope. He was concerned with selective thermal radiation, which he investigated with plates of substances that radiated and absorbed selectively for different qualities of radiation rather than maximally for all qualities of radiation. He discussed the experiments in terms of rays which could be reflected and refracted, and which obeyed the Helmholtz reciprocity principle (though he did not use an eponym for it). He did not in this paper mention that the qualities of the rays might be described by their wavelengths, nor did he use spectrally resolving apparatus such as prisms or diffraction gratings. His work was quantitative within these constraints. He made his measurements in a room temperature environment, and quickly so as to catch his bodies in a condition near the thermal equilibrium in which they had been prepared by heating to equilibrium with boiling water. His measurements confirmed that substances that emit and absorb selectively respect the principle of selective equality of emission and absorption at thermal equilibrium.

Stewart offered a theoretical proof that this should be the case separately for every selected quality of thermal radiation, but his mathematics was not rigorously valid. According to historian D.M Siegel: "He was not a practitioner of the more sophisticated techniques of nineteenth-century mathematical physics; he did not even make use of the functional notation in dealing with spectral distributions." He made no mention of thermodynamics in this paper, though he did refer to conservation of *vis viva*. He proposed that his measurements implied that radiation was both absorbed and emitted by particles of matter throughout depths of the media in which it propagated. He applied the Helmholtz reciprocity principle to account for the material interface processes as distinct from the processes in the interior material. He concluded that his experiments showed that, in the interior of an enclosure in thermal equilibrium, the radiant heat, reflected and emitted combined, leaving any part of the surface, regardless of its substance, was the same as would have left that same portion of the surface if it had been composed of lamp-black. He did not mention the possibility of ideally perfectly reflective walls; in particular he noted that highly polished real physical metals absorb very slightly.

Gustav Kirchhoff

In 1859, not knowing of Stewart's work, Gustav Robert Kirchhoff reported the coincidence of the wavelengths of spectrally resolved lines of absorption and of emission of visible light. Importantly for thermal physics, he also observed that bright lines or dark lines were apparent depending on the temperature difference between emitter and absorber.

Kirchhoff then went on to consider bodies that emit and absorb heat radiation, in an opaque enclosure or cavity, in equilibrium at temperature T.

Here is used a notation different from Kirchhoff's. Here, the emitting power $E(T, i)$ denotes a dimensioned quantity, the total radiation emitted by a body labelled by index i at temperature T. The total absorption ratio $a(T, i)$ of that body is dimensionless, the ratio of absorbed to incident radiation in the cavity at temperature T . (In contrast with Balfour Stewart's, Kirchhoff's definition of his absorption ratio did not refer in particular to a lamp-black surface as the source of the incident radiation.) Thus the ratio $E(T, i) / a(T, i)$ of emitting power to absorption ratio is a dimensioned quantity, with the dimensions of emitting power, because $a(T, i)$ is dimensionless. Also here the wavelength-specific emitting power of the body at temperature T is denoted by $E(\lambda, T, i)$ and the wavelength-specific absorption ratio by $a(\lambda, T, i)$. Again, the ratio $E(\lambda, T, i) / a(\lambda, T, i)$ of emitting power to absorption ratio is a dimensioned quantity, with the dimensions of emitting power.

In a second report made in 1859, Kirchhoff announced a new general principle or law for which he offered a theoretical and mathematical proof, though he did not offer quantitative measurements of radiation powers. His theoretical proof was and still is considered by some writers to be invalid. His principle, however, has endured: it was that for heat rays of the same wavelength, in equilibrium at a given temperature, the wavelength-specific ratio of emitting power to absorption ratio has one and the same common value for all bodies that emit and absorb at that wavelength. In symbols, the law stated that the wavelength-specific ratio $E(\lambda, T, i) / a(\lambda, T, i)$ has one and the same value for all bodies, that is for all values of index i . In this report there was no mention of black bodies.

In 1860, still not knowing of Stewart's measurements for selected qualities of radiation, Kirchhoff pointed out that it was long established experimentally that for total heat radiation, of unselected quality, emitted and absorbed by a body in equilibrium, the dimensioned total radiation ratio $E(T, i) / a(T, i)$, has one and the same value common to all bodies, that is, for every value of the material index i. Again without measurements of radiative powers or other new experimental data, Kirchhoff then offered a fresh theoretical proof of his new principle of the universality of the value of the wavelength-specific ratio $E(\lambda, T, i) / a(\lambda, T, i)$ at thermal equilibrium. His fresh theoretical proof was and still is considered by some writers to be invalid.

But more importantly, it relied on a new theoretical postulate of "perfectly black bodies", which is the reason why one speaks of Kirchhoff's law. Such black bodies showed complete absorption in their infinitely thin most superficial surface. They correspond to Balfour Stewart's reference bodies, with internal radiation, coated with lamp-black. They were not the more realistic perfectly black bodies later considered by Planck. Planck's black bodies radiated and absorbed only by the material in their interiors; their interfaces with contiguous media were only mathematical surfaces, capable neither of absorption nor emission, but only of reflecting and transmitting with refraction.

Kirchhoff's proof considered an arbitrary non-ideal body labelled i as well as various perfect black bodies labelled BB . It required that the bodies be kept in a cavity in thermal equilibrium at temperature T . His proof intended to show that the ratio $E(\lambda, T, i) / a(\lambda, T, i)$ was independent of the nature i of the non-ideal body, however partly transparent or partly reflective it was.

His proof first argued that for wavelength λ and at temperature T, at thermal equilibrium, all perfectly black bodies of the same size and shape have the one and the same common value of emissive power $E(\lambda, T, \text{BB})$, with the dimensions of power. His proof noted that the dimensionless wavelength-specific absorption ratio $a(\lambda, T, \text{BB})$ of a perfectly black body is by definition exactly 1. Then for a perfectly black body, the wavelength-specific ratio of emissive power to absorption ratio $E(\lambda, T, \text{BB}) / a(\lambda, T, \text{BB})$ is again just $E(\lambda, T, \text{BB})$, with the dimensions of power. Kirchhoff considered, successively, thermal equilibrium with the arbitrary non-ideal body, and with a perfectly black body of the same size and shape, in place in his cavity in equilibrium at temperature T . He argued that the flows of heat radiation must be the same in each case. Thus he argued that at thermal equilibrium the ratio $E(\lambda, T, i) / a(\lambda, T, i)$ was equal to $E(\lambda, T, \text{BB})$, which may now be denoted $B_\lambda(\lambda, T)$, a continuous function, dependent only on λ at fixed temperature T, and an increasing function of T at fixed wavelength λ, at low temperatures vanishing for visible but not for longer wavelengths, with positive values for visible wavelengths at higher temperatures, which does not depend on the nature i of the arbitrary non-ideal body. (Geometrical factors, taken into detailed account by Kirchhoff, have been ignored in the foregoing.)

Thus Kirchhoff's law of thermal radiation can be stated: *For any material at all, radiating and absorbing in thermodynamic equilibrium at any given temperature T, for every wavelength λ, the ratio of emissive power to absorptive ratio has one universal value, which is characteristic of a perfect black body, and is an emissive power which we here represent by $B_\lambda(\lambda, T)$* . (For our notation $B_\lambda(\lambda, T)$, Kirchhoff's original notation was simply *e*.)

Kirchhoff announced that the determination of the function $B_\lambda(\lambda, T)$ was a problem of the highest importance, though he recognized that there would be experimental difficulties to be overcome. He supposed that like other functions that do not depend on the properties of individual bodies, it would be a simple function. That function $B_{\lambda}(\lambda, T)$ has occasionally been called 'Kirchhoff's (emission, universal) function', though its precise mathematical form would not be known for another forty years, till it was discovered by Planck in 1900. The theoretical proof for Kirchhoff's universality principle was worked on and debated by various physicists over the same time, and later. Kirchhoff stated later in 1860 that his theoretical proof was better than Balfour Stewart's, and in some respects it was so. Kirchhoff's 1860 paper did not mention the second law of thermodynamics, and of course did not mention the concept of entropy which had not at that time been established. In a more considered account in a book in 1862, Kirchhoff mentioned the connection of his law with "Carnot's principle", which is a form of the second law.

According to Helge Kragh, "Quantum theory owes its origin to the study of thermal radiation, in particular to the "blackbody" radiation that Robert Kirchhoff had first defined in 1859–1860."

Empirical and Theoretical Ingredients for the Scientific Induction of Planck's Law

In 1860, Kirchhoff predicted experimental difficulties for the empirical determination of the function that described the dependence of the black-body spectrum as a function only of temperature and wavelength. And so it turned out. It took some forty years of development of improved methods of measurement of electromagnetic radiation to get a reliable result.

In 1865, John Tyndall described radiation from electrically heated filaments and from carbon arcs as visible and invisible. Tyndall spectrally decomposed the radiation by use of a rock salt prism, which passed heat as well as visible rays, and measured the radiation intensity by means of a thermopile.

In 1880, André-Prosper-Paul Crova published a diagram of the three-dimensional appearance of the graph of the strength of thermal radiation as a function of wavelength and temperature. He determined the spectral variable by use of prisms. He analyzed the surface through what he called "isothermal" curves, sections for a single temperature, with a spectral variable on the abscissa and a power variable on the ordinate. He put smooth curves through his experimental data points. They had one peak at a spectral value characteristic for the temperature, and fell either side of it towards the horizontal axis. Such spectral sections are widely shown even today.

In a series of papers from 1881 to 1886, Langley reported measurements of the spectrum of heat radiation, using diffraction gratings and prisms, and the most sensitive detectors that he could make. He reported that there was a peak intensity that increased with temperature, that the shape of the spectrum was not symmetrical about the peak, that there was a strong fall-off of intensity when the wavelength was shorter than an approximate cut-off value for each temperature, that the approximate cut-off wavelength decreased with increasing temperature, and that the wavelength of the peak intensity decreased with temperature, so that the intensity increased strongly with temperature for short wavelengths that were longer than the approximate cut-off for the temperature.

Having read Langley, in 1888, Russian physicist V.A. Michelson published a consideration of the idea that the unknown Kirchhoff radiation function could be explained physically and stated mathematically in terms of "complete irregularity of the vibrations of ... atoms". At this time, Planck was not studying radiation closely, and believed in neither atoms nor statistical physics. Michelson produced a formula for the spectrum for temperature:

$$I_\lambda = B_1\theta^{\frac{3}{2}}\exp\left(-\frac{c}{\lambda^2\theta}\right)\lambda^{-6},$$

where I_λ denotes specific radiative intensity at wavelength λ and temperature θ, and where B_1 and c are empirical constants.

In 1898, Otto Lummer and Ferdinand Kurlbaum published an account of their cavity radiation source. Their design has been used largely unchanged for radiation measurements to the present day. It was a platinum box, divided by diaphragms, with its interior blackened with iron oxide. It was an important ingredient for the progressively improved measurements that led to the discovery of Planck's law. A version described in 1901 had its interior blackened with a mixture of chromium, nickel, and cobalt oxides.

The importance of the Lummer and Kurlbaum cavity radiation source was that it was an experimentally accessible source of black-body radiation, as distinct from radiation from a simply exposed incandescent solid body, which previously was the nearest available experimental approximation to black-body radiation over a suitable range of temperatures. The simply exposed incandescent solid bodies, that had previously been used, emitted radiation with departures from the black-body spectrum that made it impossible to find the true black-body spectrum from experiments.

Planck's Views Just before the Empirical Facts Led him to Find his Eventual Law

Theoretical and empirical progress enabled Lummer and Pringsheim to write in 1899 that available experimental evidence was approximately consistent with the specific intensity law $C\lambda^{-5}e^{(-c/\lambda T)}$ where C and c denote empirically measurable constants, and where λ and T denote wavelength and temperature respectively. For theoretical reasons, Planck at that time accepted this formulation, which has an effective cut-off of short wavelengths.

Finding the Empirical Law

Max Planck originally produced his law on 19 October 1900 as an improvement upon the Wien approximation, published in 1896 by Wilhelm Wien, which fit the experimental data at short wavelengths (high frequencies) but deviated from it at long wavelengths (low frequencies).

In June 1900, based on heuristic theoretical considerations, Rayleigh had suggested a formula that he proposed might be checked experimentally. The suggestion was that the Stewart–Kirchhoff universal function might be of the form $c_1 T\lambda^{-4}\exp(-c_2/(\lambda T))$. This was not the celebrated Rayleigh–Jeans formula $8\pi k_B T\lambda^{-4}$, which did not emerge until 1905, though it did reduce to the latter for long wavelengths, which are the relevant ones here.

According to Klein, one may speculate that it is likely that Planck had seen this suggestion though he did not mention it in his papers of 1900 and 1901. Planck would have been aware of various other proposed formulas which had been offered. On 7 October 1900, Rubens told Planck that in the complementary domain (long wavelength, low frequency), and only there, Rayleigh's 1900 formula fitted the observed data well.

For long wavelengths, Rayleigh's 1900 heuristic formula approximately meant that energy was proportional to temperature, U_λ = const. T. It is known that $dS/dU_\lambda = 1/T$ and this leads to dS/dU_λ = const./ U_λ and thence to $d^2S/dU_\lambda{}^2 = -$const. / $U_\lambda{}^2$ for long wavelengths. But for short wavelengths, the Wien formula leads to $1/T = -$const. / ln U_λ + const. and thence to $d^2S/dU_\lambda{}^2 = -$ const. / U_λ for short wavelengths. Planck perhaps patched together these two heuristic formulas, for long and for short wavelengths, to produce a formula

$$d^2S/dU_\lambda^2 = \frac{\alpha}{U_\lambda(\beta + U_\lambda)}.$$

This led Planck to the formula

$$B_\lambda(T) = \frac{C\lambda^{-5}}{e^{\frac{c}{\lambda T}} - 1},$$

where Planck used the symbols C and c to denote empirical fitting constants.

Planck sent this result to Rubens, who compared it with his and Kurlbaum's observational data and found that it fitted for all wavelengths remarkably well. On 19 October 1900, Rubens and Kurlbaum briefly reported the fit to the data, and Planck added a short presentation to give a theoretical sketch to account for his formula. Within a week, Rubens and Kurlbaum gave a fuller report of their measurements confirming Planck's law. Their technique for spectral resolution of the longer wavelength radiation was called the residual ray method. The rays were repeatedly reflected from polished crystal surfaces, and the rays that made it all the way through the process were 'residual', and were of wavelengths preferentially reflected by crystals of suitably specific materials.

Trying to Find a Physical Explanation of the Law

Once Planck had discovered the empirically fitting function, he constructed a physical derivation of this law. His thinking revolved around entropy rather than being directly about temperature. Planck considered a cavity with perfectly reflective walls; the cavity contained finitely many hypothetical well separated and recognizable but identically constituted, of definite magnitude, resonant oscillatory bodies, several such oscillators at each of finitely many characteristic frequencies. The hypothetical oscillators were for Planck purely imaginary theoretical investigative probes, and he said of them that such oscillators do not need to "really exist somewhere in nature, provided their existence and their properties are consistent with the laws of thermodynamics and electrodynamics.". Planck did not attribute any definite physical significance to his hypothesis of resonant oscillators, but rather proposed it as a mathematical device that enabled him to derive a single expression for the black body spectrum that matched the empirical data at all wavelengths. He tentatively mentioned the possible connection of such oscillators with atoms. In a sense, the oscillators corresponded to Planck's speck of carbon; the size of the speck could be small regardless of the size of the cavity, provided the speck effectively transduced energy between radiative wavelength modes.

Partly following a heuristic method of calculation pioneered by Boltzmann for gas molecules, Planck considered the possible ways of distributing electromagnetic energy over the different modes of his hypothetical charged material oscillators. This acceptance of the probabilistic approach, following Boltzmann, for Planck was a radical change from his former position, which till then had deliberately opposed such thinking proposed by Boltzmann.

Heuristically, Boltzmann had distributed the energy in arbitrary merely mathematical quanta ϵ, which he had proceeded to make tend to zero in magnitude, because the finite magnitude ϕ had served only to allow definite counting for the sake of mathematical calculation of probabilities, and had no physical significance. Referring to a new universal constant of nature, *h*, Planck supposed that, in the several oscillators of each of the finitely many characteristic frequencies, the total energy was distributed to each in an integer multiple of a definite physical unit of energy, ϵ, not arbitrary as in Boltzmann's method, but now for Planck, in a new departure, characteristic of the respective characteristic frequency. His new universal constant of nature, *h*, is now known as Planck's constant.

Planck explained further that the respective definite unit, ϵ, of energy should be proportional to the respective characteristic oscillation frequency ν of the hypothetical oscillator, and in 1901 he expressed this with the constant of proportionality *h*:

$$\epsilon = h\nu.$$

Planck did not propose that light propagating in free space is quantized. The idea of quantization of the free electromagnetic field was developed later, and eventually incorporated into what we now know as quantum field theory.

In 1906 Planck acknowledged that his imaginary resonators, having linear dynamics, did not provide a physical explanation for energy transduction between frequencies. Present-day physics explains the transduction between frequencies in the presence of atoms by their quantum excitability, following Einstein. Planck believed that in a cavity with perfectly reflecting walls and with no matter present, the electromagnetic field cannot exchange energy between frequency components. This is because of the linearity of Maxwell's equations. Present-day quantum field theory predicts that, in the absence of matter, the electromagnetic field obeys nonlinear equations and in that sense does self-interact. Such interaction in the absence of matter has not yet been directly measured because it would require very high intensities and very sensitive and low-noise detectors, which are still in the process of being constructed. Planck believed that a field with no interactions neither obeys nor violates the classical principle of equipartition of energy, and instead remains exactly as it was when introduced, rather than evolving into a black body field.

Thus, the linearity of his mechanical assumptions precluded Planck from having a mechanical explanation of the maximization of the entropy of the thermodynamic equilibrium thermal radiation field. This is why he had to resort to Boltzmann's probabilistic arguments.

Planck's law may be regarded as fulfilling the prediction of Gustav Kirchhoff that his law of thermal radiation was of the highest importance. In his mature presentation of his own law, Planck offered a thorough and detailed theoretical proof for Kirchhoff's law, theoretical proof of which until then had been sometimes debated, partly because it was said to rely on unphysical theoretical objects, such as Kirchhoff's perfectly absorbing infinitely thin black surface.

Subsequent Events

It was not till five years after Planck made his heuristic assumption of abstract elements of energy or of action that Albert Einstein conceived of really existing quanta of light in 1905 as a revolutionary explanation of black-body radiation, of photoluminescence, of the photoelectric effect, and of the ionization of gases by ultraviolet light. In 1905, "Einstein believed that Planck's theory could not be made to agree with the idea of light quanta, a mistake he corrected in 1906." Contrary to Planck's beliefs of the time, Einstein proposed a model and formula whereby light was emitted, absorbed, and propagated in free space in energy quanta localized in points of space.

As an introduction to his reasoning, Einstein recapitulated Planck's model of hypothetical resonant material electric oscillators as sources and sinks of radiation, but then he offered a new argument, disconnected from that model, but partly based on a thermodynamic argument of Wien, in which Planck's formula $\phi = h\nu$ played no role. Einstein gave the energy content of such quanta in the form $R\beta\nu / N$. Thus Einstein was contradicting the undulatory theory of light held by Planck.

In 1910, criticizing a manuscript sent to him by Planck, knowing that Planck was a steady supporter of Einstein's theory of special relativity, Einstein wrote to Planck: "To me it seems absurd to have energy continuously distributed in space without assuming an aether."

According to Thomas Kuhn, it was not till 1908 that Planck more or less accepted part of Einstein's arguments for physical as distinct from abstract mathematical discreteness in thermal radiation physics. Still in 1908, considering Einstein's proposal of quantal propagation, Planck opined that such a revolutionary step was perhaps unnecessary. Until then, Planck had been consistent in thinking that discreteness of action quanta was to be found neither in his resonant oscillators nor in the propagation of thermal radiation. Kuhn wrote that, in Planck's earlier papers and in his 1906 monograph, there is no "mention of discontinuity, [nor] of talk of a restriction on oscillator energy, [nor of] any formula like $U = nh\nu$." Kuhn pointed out that his study of Planck's papers of 1900 and 1901, and of his monograph of 1906, had led him to "heretical" conclusions, contrary to the widespread assumptions of others who saw Planck's writing only from the perspective of later, anachronistic, viewpoints. Kuhn's conclusions, finding a period till 1908, when Planck consistently held his 'first theory', have been accepted by other historians.

In the second edition of his monograph, in 1912, Planck sustained his dissent from Einstein's proposal of light quanta. He proposed in some detail that absorption of light by his virtual material resonators might be continuous, occurring at a constant rate in equilibrium, as distinct from quantal absorption. Only emission was quantal. This has at times been called Planck's "second theory".

It was not till 1919 that Planck in the third edition of his monograph more or less accepted his 'third theory', that both emission and absorption of light were quantal.

The colourful term "ultraviolet catastrophe" was given by Paul Ehrenfest in 1911 to the paradoxical result that the total energy in the cavity tends to infinity when the equipartition theorem of classical statistical mechanics is (mistakenly) applied to black body radiation. But this had not been part of Planck's thinking, because he had not tried to apply the doctrine of equipartition: when he made his discovery in 1900, he had not noticed any sort of "catastrophe". It was first noted by Lord Rayleigh in 1900, and then in 1901 by Sir James Jeans; and later, in 1905, by Einstein when he wanted to support the idea that light propagates as discrete packets, later called 'photons', and by Rayleigh and by Jeans.

In 1913, Bohr gave another formula with a further different physical meaning to the quantity $h\nu$. In contrast to Planck's and Einstein's formulas, Bohr's formula referred explicitly and categorically to energy levels of atoms. Bohr's formula was $W_{\tau_2} - W_{\tau_1} = h\nu$ where W_{τ_2} and W_{τ_1} denote the energy levels of quantum states of an atom, with quantum numbers τ_2 and τ_1 .

The symbol ν denotes the frequency of a quantum of radiation that can be emitted or absorbed as the atom passes between those two quantum states. In contrast to Planck's model, the frequency has no immediate relation to frequencies that might describe those quantum states themselves.

Later, in 1924, Satyendra Nath Bose developed the theory of the statistical mechanics of photons, which allowed a theoretical derivation of Planck's law. The actual word 'photon' was invented still later, by G.N. Lewis in 1926, who mistakenly believed that photons were conserved, contrary to Bose–Einstein statistics; nevertheless the word 'photon' was adopted to express the Einstein postulate of the packet nature of light propagation. In an electromagnetic field isolated in a vacuum in a vessel with perfectly reflective walls, such as was considered by Planck, indeed the photons would be conserved according to Einstein's 1905 model, but Lewis was referring to a field of photons considered as a system closed with respect to ponderable matter but open to exchange of electromagnetic energy with a surrounding system of ponderable matter, and he mistakenly imagined that still the photons were conserved, being stored inside atoms.

Ultimately, Planck's law of black-body radiation contributed to Einstein's concept of quanta of light carrying linear momentum, which became the fundamental basis for the development of quantum mechanics.

The above-mentioned linearity of Planck's mechanical assumptions, not allowing for energetic interactions between frequency components, was superseded in 1925 by Heisenberg's original quantum mechanics. In his paper submitted on 29 July 1925, Heisenberg's theory accounted for Bohr's above-mentioned formula of 1913. It admitted non-linear oscillators as models of atomic quantum states, allowing energetic interaction between their own multiple internal discrete Fourier frequency components, on the occasions of emission or absorption of quanta of radiation. The frequency of a quantum of radiation was that of a definite coupling between internal atomic meta-stable oscillatory quantum states. At that time, Heisenberg knew nothing of matrix algebra, but Max Born read the manuscript of Heisenberg's paper and recognized the matrix character of Heisenberg's theory. Then Born and Jordan published an explicitly matrix theory of quantum mechanics, based on, but in form distinctly different from, Heisenberg's original quantum mechanics; it is the Born and Jordan matrix theory that is today called matrix mechanics. Heisenberg's explanation of the Planck oscillators, as non-linear effects apparent as Fourier modes of transient processes of emission or absorption of radiation, showed why Planck's oscillators, viewed as enduring physical objects such as might be envisaged by classical physics, did not give an adequate explanation of the phenomena.

Nowadays, as a statement of the energy of a light quantum, often one finds the formula $E = \hbar\omega$, where $\hbar = h/2\pi$, and $\omega = 2\pi\nu$ denotes angular frequency, and less often the equivalent formula $E = h\nu$. This statement about a really existing and propagating light quantum, based on Einstein's, has a physical meaning different from that of Planck's above statement $\epsilon = h\nu$ about the abstract energy units to be distributed amongst his hypothetical resonant material oscillators.

An article by Helge Kragh published in *Physics World* gives an account of this history.

Wien's Displacement Law

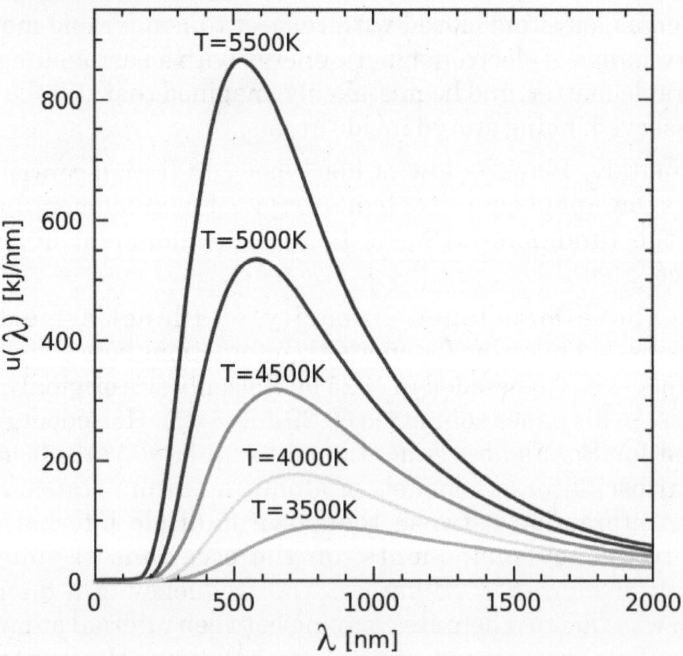

Figure: *Black body radiation as a function of wavelength for various absolute temperatures. Each curve is seen to peak at a somewhat different wavelength; Wien's law describes the shift of that peak in terms of temperature.*

Wien's displacement law states that the black body radiation curve for different temperatures peaks at a wavelength inversely proportional to the temperature. The shift of that peak is a direct consequence of the Planck radiation law which describes the spectral brightness of black body radiation as a function of wavelength at any given temperature. However it had been discovered by Wilhelm Wien several years before Max Planck developed that more general equation, and describes the entire shift of the spectrum of black body radiation toward shorter wavelengths as temperature increases.

Formally, Wien's displacement law states that the spectral radiance of black body radiation per unit wavelength, peaks at the wavelength λ_{max} given by:

$$\lambda_{\max} = \frac{b}{T}$$

where T is the absolute temperature in degrees kelvin. b is a constant of proportionality called *Wien's displacement constant*, equal to $2.8977721(26)\times10^{-3}$ m K., or more conveniently to obtain wavelength in microns, $b \approx 2900$ μm K. If one is considering the peak of black body emission per unit frequency or per proportional bandwidth, one must use a different proportionality constant. However the form of the law remains the same: the peak wavelength is inversely proportional to temperature (or the peak frequency is directly proportional to temperature).

Wien's displacement law may be referred to as "Wien's law", a term which is also used for the Wien approximation.

Examples

Wien's displacement law is relevant to some everyday experiences:

- A piece of metal heated by a torch first becomes "red hot" as the very longest visible wavelengths appear red, then becomes more orange-red as the temperature is increased, and at very high temperatures would be described as "white hot" as shorter and shorter wavelengths come to predominate the black body emission spectrum. Before it had even reached the red hot temperature, the thermal emission was mainly at longer infrared wavelengths which are not visible; nevertheless that radiation could be felt as it warms one's nearby skin.
- One easily observes changes in the colour of an incandescent light bulb (which produces light through black body radiation) as the temperature of its filament is varied by a light dimmer. As the light is dimmed and the filament temperature decreases,

the distribution of colour shifts toward longer wavelengths and the light appears redder, as well as dimmer.

- It is easy to calculate that a wood fire at 1500° K puts out peak radiation at about 2000nm. 98% of its radiation is beyond 1000nm and only a tiny proportion at visible wavelengths. Consequently, a campfire can keep one warm but is a poor source of visible light.
- The effective temperature of the Sun is 5778° K. Using Wien's law, one finds a peak emission per nanometre (of wavelength) at a wavelength of about 500 nm in the green portion of the spectrum near the peak sensitivity of the human eye. On the other hand, in terms of power per unit optical frequency, the Sun's peak emission is at 343 THz or a wavelength of 883 nm in the near infrared. In terms of power per percentage bandwidth, the peak is at about 635 nm, a red wavelength. Regardless of how one wants to plot the spectrum, about half of the sun's radiation is at wavelengths shorter than 710 nm, about the limit of the human vision. Of that, about 12% is at wavelengths shorter than 400 nm, ultraviolet wavelengths which cannot be seen. It can be appreciated that a rather large amount of the Sun's radiation falls in the fairly small visible spectrum.
- The preponderance of emission in the visible range, however, is not the case in most stars. The hot supergiant Rigel emits 60% of its light in the ultraviolet, while the cool supergiant Betelgeuse emits 85% of its light at infrared wavelengths. With both stars prominent in the constellation of Orion, one can easily appreciate the colour difference between the blue-white Rigel (T=12100° K) and the red Betelgeuse (T≈3300° K). While few stars are as hot as Rigel, stars cooler than the sun or even as cool as Betelgeuse are very commonplace.
- Mammals with a skin temperature of about 300° K emit peak radiation at around 10 μm in the far infrared. This is therefore the range of infrared wavelengths that pit viper snakes and passive IR cameras must sense.
- When comparing the apparent colour of lighting sources (including fluorescent lights, LED lighting, computer monitors, and photoflash), it is customary to cite the colour temperature. Although the spectra of such lights are not accurately described by the black body radiation curve, a colour temperature is quoted for which black body radiation would most closely match the

subjective colour of that source. For instance, the blue-white fluorescent light sometimes used in an office may have a colour temperature of 6500° K, whereas the reddish tint of a dimmed incandescent light may have a colour temperature (and an actual filament temperature) of 2000° K.

Note that the informal description of the former (bluish) colour as "cool" and the latter (reddish) as "warm" is exactly opposite the actual temperature change involved in black body radiation!

Discovery

The law is named for Wilhelm Wien, who derived it in 1893 based on a thermodynamic argument. Wien considered adiabatic expansion of a cavity containing waves of light in thermal equilibrium. He showed that under slow expansion or contraction, the energy of light reflecting off the walls changes in exactly the same way as the frequency. A general principle of thermodynamics is that a thermal equilibrium state, when expanded very slowly stays in thermal equilibrium. The adiabatic principle allowed Wien to conclude that for each mode, the adiabatic invariant energy/frequency is only a function of the other adiabatic invariant, the frequency/temperature. A modern variant of Wien's derivation can be found in the textbook by Wannier.

The consequence is that the shape of the black body radiation function (which was not yet understood) would shift proportionally in frequency (or inversely proportionally in wavelength) with temperature. When Max Planck later formulated the correct black body radiation function it did not include Wien's constant *b* explicitly. Rather, Planck's constant *h* was created and introduced into his new formula. From Planck's constant *h* and the Boltzmann constant *k*, Wien's constant *b* can be obtained.

Frequency-dependent Formulation

In terms of frequency υ (in hertz), Wien's displacement law predicts a peak emission *per unit frequency* at ν_{max} given by:

$$\nu_{max} = \frac{\alpha}{h} kT \approx (5.879 \times 10^{10} \text{ Hz/K}) \cdot T$$

where $\alpha \approx 2.821439...$ is a constant resulting from the numerical solution of the maximization equation, *k* is the Boltzmann constant, *h* is the Planck constant, and *T* is the temperature (in kelvins). This frequency, it must be remembered, does *not* correspond to the wavelength from the earlier formula which considered the peak emission *per unit wavelength.*

Derivation from Planck's Law

Planck's law for the spectrum of black body radiation predicts the Wien displacement law and may be used to numerically evaluate the constant relating temperature and peak wavelength (or frequency). According to one form of that law, the black body spectral radiance (power per emitting area per solid angle *per unit wavelength*) is given by:

$$u_\lambda(\lambda,T)=\frac{2hc^2}{\lambda^5}\frac{1}{e^{hc/\lambda kT}-1}.$$

Differentiating $u(\lambda,T)$ with respect to λ and setting the derivative equal to zero gives

$$\frac{\partial u}{\partial\lambda}=2hc^2\left(\frac{hc}{kT\lambda^7}\frac{e^{hc/\lambda kT}}{\left(e^{hc/\lambda kT}-1\right)^2}-\frac{1}{\lambda^6}\frac{5}{e^{hc/\lambda kT}-1}\right)=0,$$

which can be simplified to give

$$\frac{hc}{\lambda kT}\frac{e^{hc/\lambda kT}}{e^{hc/\lambda kT}-1}-5=0.$$

By defining:

$$x\equiv\frac{hc}{\lambda kT},$$

the equation becomes one in that single variable:

$$\frac{xe^x}{e^x-1}-5=0.$$

The numerical solution to this equation is x = 4.965114231.

Solving for the wavelength λ in units of manometres, and using kelvins for the temperature yields:

$$\lambda_{\max}=\frac{hc}{x}\frac{1}{kT}=\frac{2.89776829\times10^6\ \text{nm}\cdot K}{T}.$$

The form of Wien's displacement law in terms of maximum radiance per unit *frequency* is derived using similar methods, but starting with the form of Planck's law expressed in those terms rather than wavelength. The effective result is to substitute 3 for 5 in the equation for the peak wavelength. This is solved with $x = 2.821439372$.

Using the value *4* in this equation solves for the wavelength of the peak in the spectral radiance expressed in radiance *per proportional bandwidth*, perhaps a fairer way of presenting "wavelength of peak emission." That is solved as $x = 3.920690395$. The important point of

Wiens law, however, is that *any* of these wavelength markers, including the median wavelength (or the wavelength below which P% of the emission takes place, for instance) scales according to the reciprocal of temperature.

Kirchhoff's Circuit Laws

Kirchhoff's circuit laws are two equalities that deal with the current and potential difference (commonly known as voltage) in the lumped element model of electrical circuits. They were first described in 1845 by German physicist Gustav Kirchhoff. This generalized the work of Georg Ohm and preceded the work of Maxwell. Widely used in electrical engineering, they are also called Kirchhoff's rules or simply Kirchhoff's laws.

Both of Kirchhoff's laws can be understood as corollaries of the Maxwell equations in the low-frequency limit. They are accurate for DC circuits, and for AC circuits at frequencies where the wavelengths of electromagnetic radiation are very large compared to the circuits.

Kirchhoff's Current Law (KCL)

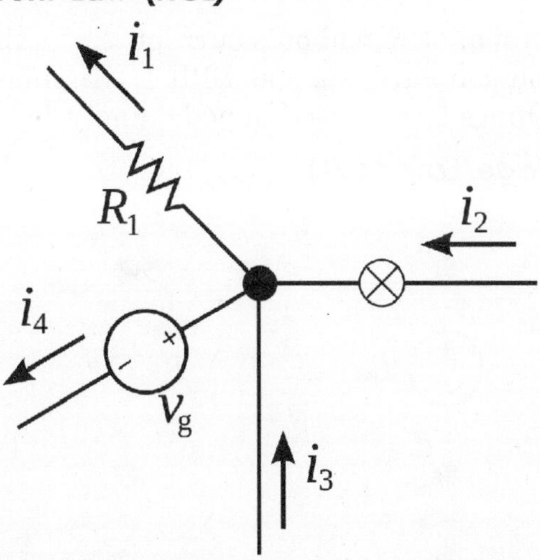

Figure: *The current entering any junction is equal to the current leaving that junction.* $i_2 + i_3 = i_1 + i_4$

This law is also called Kirchhoff's first law, Kirchhoff's point rule, or Kirchhoff's junction rule (or nodal rule).

The principle of conservation of electric charge implies that:

At any node (junction) in an electrical circuit, the sum of currents flowing into that node is equal to the sum of currents flowing out of that node or equivalently

The algebraic sum of currents in a network of conductors meeting at a point is zero.

Recalling that current is a signed (positive or negative) quantity reflecting direction towards or away from a node, this principle can be stated as:

$$\sum_{k=1}^{n} I_k = 0$$

n is the total number of branches with currents flowing towards or away from the node.

This formula is valid for complex currents:

$$\sum_{k=1}^{n} \tilde{I}_k = 0$$

The law is based on the conservation of charge whereby the charge (measured in coulombs) is the product of the current (in amperes) and the time (in seconds).

Uses

A matrix version of Kirchhoff's current law is the basis of most circuit simulation software, such as SPICE. Kirchhoff's current law combined with Ohm's Law is used in nodal analysis.

Kirchhoff's Voltage Law (KVL)

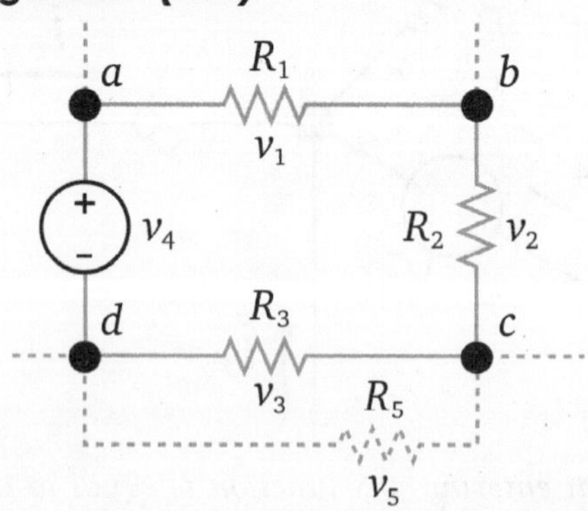

Figure: *The sum of all the voltages around the loop is equal to zero.*

$$v_1 + v_2 + v_3 - v_4 = 0$$

This law is also called Kirchhoff's second law, Kirchhoff's loop (or mesh) rule, and Kirchhoff's second rule.

The principle of conservation of energy implies that

The directed sum of the electrical potential differences (voltage) around any closed network is zero, or:

More simply, the sum of the emfs in any closed loop is equivalent to the sum of the potential drops in that loop, or:

The algebraic sum of the products of the resistances of the conductors and the currents in them in a closed loop is equal to the total emf available in that loop.

Similarly to KCL, it can be stated as:

$$\sum_{k=1}^{n} V_k = 0$$

Here, n is the total number of voltages measured. The voltages may also be complex:

$$\sum_{k=1}^{n} \tilde{V}_k = 0$$

This law is based on the conservation of energy whereby voltage is defined as the energy per unit charge. The total amount of energy gained per unit charge must be equal to the amount of energy lost per unit charge, as energy and charge are both conserved.

Generalization

In the low-frequency limit, the voltage drop around any loop is zero. This includes imaginary loops arranged arbitrarily in space – not limited to the loops delineated by the circuit elements and conductors. In the low-frequency limit, this is a corollary of Faraday's law of induction (which is one of the Maxwell equations).

This has practical application in situations involving "static electricity".

Limitations

KCL and KVL both depend on the lumped element model being applicable to the circuit in question. When the model is not applicable, the laws do not apply.

KCL, in its usual form, is dependent on the assumption that current flows only in conductors, and that whenever current flows into one end of a conductor it immediately flows out the other end. This is not a safe assumption for high-frequency AC circuits, where the lumped element model is no longer applicable.

It is often possible to improve the applicability of KCL by considering "parasitic capacitances" distributed along the conductors. Significant violations of KCL can occur even at 60Hz, which is not a very high frequency.

In other words, KCL is valid only if the total electric charge, Q, remains constant in the region being considered. In practical cases this is always so when KCL is applied at a geometric point. When investigating a finite region, however, it is possible that the charge density within the region may change. Since charge is conserved, this can only come about by a flow of charge across the region boundary. This flow represents a net current, and KCL is violated.

KVL is based on the assumption that there is no fluctuating magnetic field linking the closed loop. This is not a safe assumption for high-frequency (short-wavelength) AC circuits. In the presence of a changing magnetic field the electric field is not a conservative vector field. Therefore the electric field cannot be the gradient of any potential. That is to say, the line integral of the electric field around the loop is not zero, directly contradicting KVL.

It is often possible to improve the applicability of KVL by considering "parasitic inductances" (including mutual inductances) distributed along the conductors. These are treated as imaginary circuit elements that produce a voltage drop equal to the rate-of-change of the flux.

Example

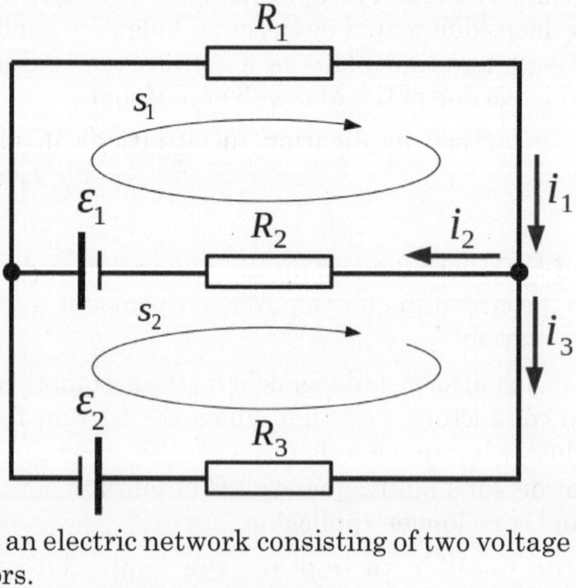

Assume an electric network consisting of two voltage sources and three resistors.

According to the first law we have

$$i_1 - i_2 - i_3 = 0$$

The second law applied to the closed circuit s_1 gives

$$-R_2 i_2 + \epsilon_1 - R_1 i_1 = 0$$

The second law applied to the closed circuit s_2 gives

$$-R_3 i_3 - \epsilon_2 - \epsilon_1 + R_2 i_2 = 0$$

Thus we get a linear system of equations in i_1, i_2, i_3:

$$\begin{cases} i_1 - i_2 - i_3 & = 0 \\ -R_2 i_2 + \epsilon_1 - R_1 i_1 & = 0 \\ -R_3 i_3 - \epsilon_2 - \epsilon_1 + R_2 i_2 & = 0 \end{cases}$$

Assuming

$$R_1 = 100, R_2 = 200, R_3 = 300 \text{ (ohms)}; \epsilon_1 = 3, \epsilon_2 = 4 \text{ (volts)}$$

the solution is

$$\begin{cases} i_1 = \dfrac{1}{1100} \\ i_2 = \dfrac{4}{275} \\ i_3 = -\dfrac{3}{220} \end{cases}$$

i_3 has a negative sign, which means that the direction of is opposite to the assumed direction (the direction defined in the picture).

Chapter 8

Solar Constant

The solar constant, a measure of flux density, is the amount of solar electromagnetic radiation (the solar irradiance) per unit area that would be incident on a plane perpendicular to the rays, at a distance of one astronomical unit (AU) from the Sun (roughly the mean distance from the Sun to the Earth).

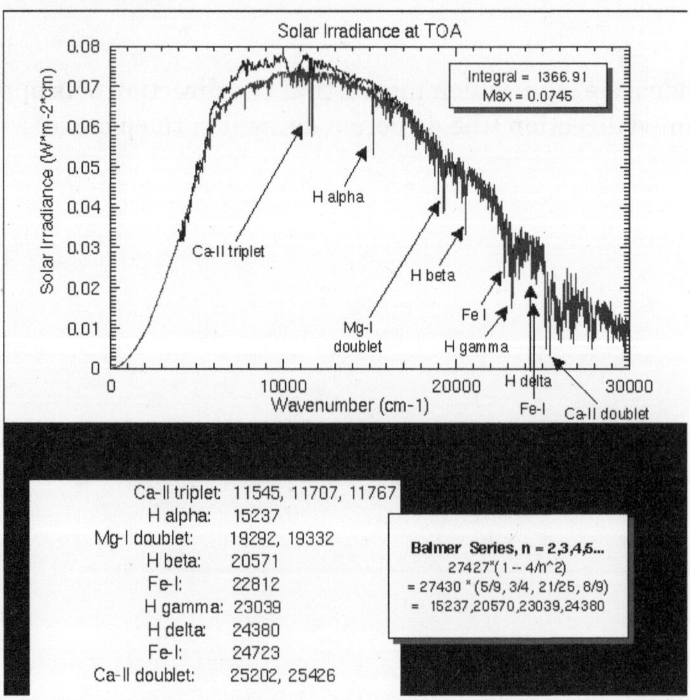

***Figure:** Solar irradiance spectrum at top of atmosphere, on a linear scale and plotted against wavenumber.*

The solar constant includes all types of solar radiation, not just the visible light. It is measured by satellite to be roughly 1.361 kilowatts per square metre (kW/m^2) at solar minimum and approximately 0.1% greater (roughly 1.362 kW/m^2) at solar maximum.

Calculation

Solar irradiance is measured by satellite near the outer surface of Earth's atmosphere, and is then adjusted using the inverse square law to infer the magnitude of solar irradiance at one AU and find the solar constant. The approximate average value cited, 1.361 kW/m^2, which is 81.66 kJ/m^2 per minute, is equivalent to approximately 1.952 calories per minute per square centimetre, or 1.952 langleys per minute.

Solar output is nearly, but not quite, constant. Variations in total solar irradiance were too small to detect with technology available before the satellite era. Total solar output is now measured as varying (over the last three 11-year sunspot cycles) by approximately 0.1%.

Historical Measurements

In 1838, Claude Pouillet made the first estimate of the solar constant. Using a very simple pyrheliometre he developed, he obtained a value of 1.228 kW/m^2, very close to the current estimate.

In 1875, Jules Violle resumed the work of Pouillet and offered a somewhat larger estimate of 1.7 kW/m^2 based, in part, on a celebrated measurement that he made from Mont Blanc in France.

In 1884, Samuel Pierpont Langley attempted to estimate the solar constant from Mount Whitney in California. By taking readings at different times of day, he tried to correct for effects due to atmospheric absorption. However, the final value he proposed, 2.903 kW/m^2, was much too large.

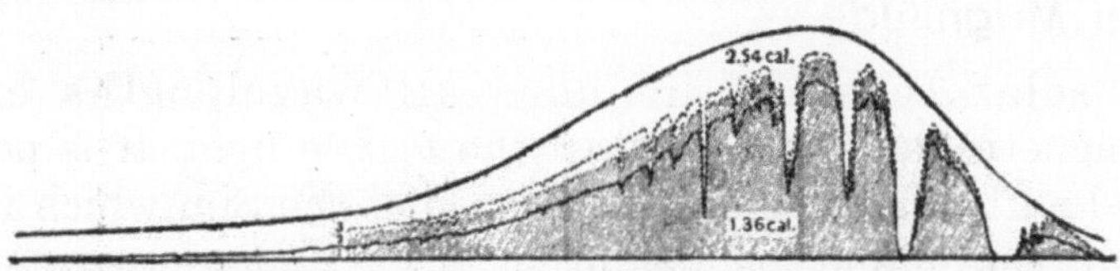

Figure: *A 1903 Langley bolograph with an erroneous solar constant of 2.54 calories/minute/square centimetre.*

Between 1902 and 1957, measurements by Charles Greeley Abbot and others at various high-altitude sites found values between 1.322 and 1.465 kW/m^2. Abbot showed that one of Langley's corrections was erroneously applied. Abbot's results varied between 1.89 and 2.22 calories (1.318 to 1.548 kW/m^2), a variation that appeared to be due to the Sun and not the Earth's atmosphere.

Relationship to Other Measurements

Solar Irradiance

The actual direct solar irradiance at the top of the atmosphere fluctuates by about 6.9% during a year (from 1.412 kW/m^2 in early January to 1.321 kW/m^2 in early July) due to the Earth's varying distance from the Sun, and typically by much less than 0.1% from day to day. Thus, for the whole Earth (which has a cross section of 127,400,000 km^2), the power is 1.730×10^{17} W (or 173,000 terawatts), plus or minus 2 W/m^2.

The solar constant does not remain constant over long periods of time, but over a year the solar constant varies much less than the solar irradiance measured at the top of the atmosphere. This is because the solar constant is measured at a fixed distance of 1 AU while the solar irradiance will be affected by the ellipticity of the Earth's orbit.

The Earth receives a total amount of radiation determined by its cross section ($\pi \cdot R_E^2$), but as it rotates this energy is distributed across the entire surface area ($4 \pi \cdot R_E^2$).

Hence the average incoming solar radiation, taking into account the angle at which the rays strike and that at any one moment half the planet does not receive any solar radiation, is one-fourth the solar constant (approximately 340 W/m^2).

The amount reaching the Earth's surface (as insolation) is further reduced by atmospheric attenutation, which varies. At any given moment, the amount of solar radiation received at a location on the Earth's surface depends on the state of the atmosphere, the location's latitude, and the time of day.

Apparent Magnitude

The solar constant includes all wavelengths of solar electromagnetic radiation, not just the visible light. It is positively correlated with the apparent magnitude of the Sun which is –26.8. The solar constant and the magnitude of the Sun are two methods of describing the apparent brightness of the Sun, though the magnitude is based on the Sun's visual output only.

The Sun's Total Radiation

The angular diametre of the Earth as seen from the Sun is approximately 1/11,700 radians (about 18 arc-seconds), meaning the solid angle of the Earth as seen from the Sun is approximately 1/175,000,000 of a steradian. Thus the Sun emits about 2.2 billion times the amount of radiation that is caught by Earth, in other words about 3.86×10^{26} watts.

Albedo

The term albino is usually defined as the fraction or percent of the reflected solar radiation from the surface to incoming solar radiation.

Thus, it is usually refers to the reflectivity of a particular band or portion of the spectrum.The albedo of the whole earth and atmospheric system approximates 35%

Albino is also defined as the ratio of reflected radiations to the total incident radiation.

Greenhouse Effect

The greenhouse effect is a process by which thermal radiation from a planetary surface is absorbed by atmospheric greenhouse gases, and is re-radiated in all directions. Since part of this re-radiation is back towards the surface and the lower atmosphere, it results in an elevation of the average surface temperature above what it would be in the absence of the gases.

Solar radiation at the frequencies of visible light largely passes through the atmosphere to warm the planetary surface, which then emits this energy at the lower frequencies of infrared thermal radiation. Infrared radiation is absorbed by greenhouse gases, which in turn re-radiate much of the energy to the surface and lower atmosphere. The mechanism is named after the effect of solar radiation passing through glass and warming a greenhouse, but the way it retains heat is fundamentally different as a greenhouse works by reducing airflow, isolating the warm air inside the structure so that heat is not lost by convection.

If an ideal thermally conductive blackbody were the same distance from the Sun as the Earth is, it would have a temperature of about 5.3°C. However, since the Earth reflects about 30% of the incoming sunlight, this idealized planet's effective temperature (the temperature of a blackbody that would emit the same amount of radiation) would be about –18°C.

The surface temperature of this hypothetical planet is 33°C below Earth's actual surface temperature of approximately 14°C. The mechanism that produces this difference between the actual surface temperature and the effective temperature is due to the atmosphere and is known as the greenhouse effect.

Earth's natural greenhouse effect makes life as we know it possible. However, human activities, primarily the burning of fossil fuels and clearing of forests, have intensified the natural greenhouse effect, causing global warming.

History

The existence of the greenhouse effect was argued for by Joseph Fourier in 1824. The argument and the evidence was further strengthened by Claude Pouillet in 1827 and 1838, and reasoned from experimental observations by John Tyndall in 1859, and more fully quantified by Svante Arrhenius in 1896.

In 1917 Alexander Graham Bell wrote "[The unchecked burning of fossil fuels] would have a sort of greenhouse effect", and "The net result is the greenhouse becomes a sort of hot-house." Bell went on to also advocate for the use of alternate energy sources, such as solar energy.

Mechanism

The Earth receives energy from the Sun in the form UV, visible, and near IR radiation, most of which passes through the atmosphere without being absorbed. Of the total amount of energy available at the top of the atmosphere (TOA), about 50% is absorbed at the Earth's surface. Because it is warm, the surface radiates far IR thermal radiation that consists of wavelengths that are predominantly much longer than the wavelengths that were absorbed (the overlap between the incident solar spectrum and the terrestrial thermal spectrum is small enough to be neglected for most purposes).

Most of this thermal radiation is absorbed by the atmosphere and re-radiated both upwards and downwards; that radiated downwards is absorbed by the Earth's surface. This trapping of long-wavelength thermal radiation leads to a higher equilibrium temperature than if the atmosphere were absent.

This highly simplified picture of the basic mechanism needs to be qualified in a number of ways, none of which affect the fundamental process.

- The incoming radiation from the Sun is mostly in the form of visible light and nearby wavelengths, largely in the range 0.2–4 μm, corresponding to the Sun's radiative temperature of 6,000 K. Almost half the radiation is in the form of "visible" light, which our eyes are adapted to use.

- About 50% of the Sun's energy is absorbed at the Earth's surface and the rest is reflected or absorbed by the atmosphere. The reflection of light back into space—largely by clouds—does not much affect the basic mechanism; this light, effectively, is lost to the system.
- The absorbed energy warms the surface. Simple presentations of the greenhouse effect, such as the idealized greenhouse model, show this heat being lost as thermal radiation. The reality is more complex: the atmosphere near the surface is largely opaque to thermal radiation (with important exceptions for "window" bands), and most heat loss from the surface is by sensible heat and latent heat transport.

  Radiative energy losses become increasingly important higher in the atmosphere largely because of the decreasing concentration of water vapour, an important greenhouse gas. It is more realistic to think of the greenhouse effect as applying to a "surface" in the mid-troposphere, which is effectively coupled to the surface by a lapse rate.
- The simple picture assumes a steady state. In the real world there is the diurnal cycle as well as seasonal cycles and weather. Solar heating only applies during daytime. During the night, the atmosphere cools somewhat, but not greatly, because its emissivity is low, and during the day the atmosphere warms. Diurnal temperature changes decrease with height in the atmosphere.
- Within the region where radiative effects are important the description given by the idealized greenhouse model becomes realistic: The surface of the Earth, warmed to a temperature around 255 K, radiates long-wavelength, infrared heat in the range 4–100 μm. At these wavelengths, greenhouse gases that were largely transparent to incoming solar radiation are more absorbent. Each layer of atmosphere with greenhouses gases absorbs some of the heat being radiated upwards from lower layers. It re-radiates in all directions, both upwards and downwards; in equilibrium (by definition) the same amount as it has absorbed. This results in more warmth below. Increasing the concentration of the gases increases the amount of absorption and re-radiation, and thereby further warms the layers and ultimately the surface below.
- Greenhouse gases—including most diatomic gases with two different atoms (such as carbon monoxide, CO) and all gases

with three or more atoms—are able to absorb and emit infrared radiation. Though more than 99% of the dry atmosphere is IR transparent (because the main constituents—N_2, O_2, and Ar—are not able to directly absorb or emit infrared radiation), intermolecular collisions cause the energy absorbed and emitted by the greenhouse gases to be shared with the other, non-IR-active, gases.

Greenhouse Gases

By their percentage contribution to the greenhouse effect on Earth the four major gases are:

- water vapour, 36–70%
- carbon dioxide, 9–26%
- methane, 4–9%
- ozone, 3–7%

The major non-gas contributor to the Earth's greenhouse effect, clouds, also absorb and emit infrared radiation and thus have an effect on radiative properties of the atmosphere.

Role in Climate Change

Atmospheric gases only absorb some wavelengths of energy but are transparent to others. The absorption patterns of water vapour (blue peaks) and carbon dioxide (pink peaks) overlap in some wavelengths. Carbon dioxide is not as strong a greenhouse gas as water vapour, but it absorbs energy in wavelengths (12-15 micrometres) that water vapour does not, partially closing the "window" through which heat radiated by the surface would normally escape to space.

Strengthening of the greenhouse effect through human activities is known as the enhanced (or anthropogenic) greenhouse effect. This increase in radiative forcing from human activity is attributable mainly to increased atmospheric carbon dioxide levels. According to the latest Assessment Report from the Intergovernmental Panel on Climate Change, "*most of the observed increase in globally averaged temperatures since the mid-20th century is very likely due to the observed increase in anthropogenic greenhouse gas concentrations*".

CO_2 is produced by fossil fuel burning and other activities such as cement production and tropical deforestation. Measurements of CO_2 from the Mauna Loa observatory show that concentrations have increased from about 313 ppm in 1960 to about 389 ppm in 2010. It reached the 400ppm milestone on May 9, 2013. The current observed amount of CO_2 exceeds the geological record maxima (~300 ppm) from ice core data. The effect of combustion-produced carbon dioxide on the global climate, a special case of the greenhouse effect first described in 1896 by Svante Arrhenius, has also been called the Callendar effect.

Over the past 800,000 years, ice core data shows that carbon dioxide has varied from values as low as 180 parts per million (ppm) to the pre-industrial level of 270ppm. Paleoclimatologists consider variations in carbon dioxide concentration to be a fundamental factor influencing climate variations over this time scale.

Real Greenhouses

The "greenhouse effect" of the atmosphere is named by analogy to greenhouses which get warmer in sunlight, but the mechanism by which the atmosphere retains heat is different. A greenhouse works primarily by allowing sunlight to warm surfaces inside the structure, but then preventing absorbed heat from leaving the structure through convection, i.e. sensible heat transport. The "greenhouse effect" heats the Earth because greenhouse gases absorb outgoing radiative energy, heating the atmosphere which then emits radiative energy with some of it going back towards the Earth.

A greenhouse is built of any material that passes sunlight, usually glass, or plastic. It mainly heats up because the Sun warms the ground inside, which then warms the air in the greenhouse. The air continues to heat because it is confined within the greenhouse, unlike the environment outside the greenhouse where warm air near the surface rises and mixes with cooler air aloft. This can be demonstrated by opening a small window near the roof of a greenhouse: the temperature will drop considerably. It has also been demonstrated experimentally (R. W. Wood, 1909) that a "greenhouse" with a cover of rock salt (which is transparent to infra red) heats up an enclosure similarly to one with a glass cover. Thus greenhouses work primarily by preventing convective cooling.

In contrast, the greenhouse effect heats the Earth because rather than retaining (sensible) heat by physically preventing movement of the air, greenhouse gases act to warm the Earth by re-radiating some of the energy back towards the surface. This process may exist in real greenhouses, but is comparatively unimportant there.

Bodies Other than Earth

In the Solar System, Mars, Venus, and the moon Titan also exhibit greenhouse effects; that on Venus is particularly large, due to its atmosphere, which consists mainly of dense carbon dioxide. Titan has an anti-greenhouse effect, in that its atmosphere absorbs solar radiation but is relatively transparent to infrared radiation. Pluto also exhibits behaviour superficially similar to the anti-greenhouse effect.

A runaway greenhouse effect occurs if positive feedbacks lead to the evaporation of all greenhouse gases into the atmosphere. A runaway greenhouse effect involving carbon dioxide and water vapour is thought to have occurred on Venus.

Radiation Balance or Net Radiation

The net radiation is the difference between the total downward and upward radiation fluxes and is a measure of the energy available at the ground surface. The balance of energy after gain and loss of both short wave and long wave radiation fluxes is known as net radiation.

Net radiation represents the amount of energy, which is used for various kinds of activities. It is dispensed as sensible heat, latent heat and also in physiological processes such as photosynthesis and respiration.

The importance of this parametre is that it is the fundamental quantity of energy available at the earth's surface to drive he processes of evaporation, air and soil heat fluxes as well as other smaller energy consuming processes like photosynthesis etc.

If we consider the extra terrestrial radiation reaching annually (338wm-2)as 100%; Then out of this-

The net radiation reaching (SW) =100-28+25)-47% =114%
the earth surface.

Long wave radiation reaching at earth surface =+96%

Therefore, Net long wave radiation =-114+96 = -18%

Net all wave radiation at earth surface =+47-18 = 29%

This surplus energy is used at the earth surface for

a) Sensible heat (QH) =4%

b) Latent heat (Eva)(QE)=25%

29%

Factors Affecting Solar Radiations

The amount of insulation received at particular place and time depends on the following factors:

1. Distance from the sun.
2. Duration of daily sunlight period.
3. Solar elevation or inclinations of the solar rays to the horizon.
4. Transparency of the atmosphere towards heat radiation and
5. Output of solar radiation.

The first three of these reasons are intimately connected with revolution of earth, It is to be noted here that the earth revolves about the sun in elliptic orbit and makes one complete revolution in 365 days, simultaneously it spins about itself and complete one rotation in 24 hrs.

The average distance of the earth from the sun is 149.5 million km).

The duration f daylight also varies with the latitude and season. Longer the day light duration, greater is the insulation received, In the solar region the duration of day light is 24 hrs during summer and minimum of zero in winter season.

Chapter 9

Significance of Radiation in Agriculture

The importance of the radiation in crop production is as follow:

1. It provides the necessary energy for all the phenomena concerning biomass production.
2. Photo synthetically Active Radiations (PAR) are the real source of energy for photosynthesis process. Plants are the efficient biological converters of solar energy into biomass. Radiation defines the yield of crop in particular region.
3. It laso provides the energy for the physical processes taking place in plants, soil and atmosphere.
4. It conditions the distribution of temperature and hence crop distribution on the earth surface.

Atmospheric Temperature

Definition

The degree of hotness is known as temperature increases. Temperature is a fundamental elopement of climate from many points of view, the most important in controlling the distribution of life on the earth. Most of the weather elements are dependent on it, directly or indirectly. Air of atmosphere receives the heat energy from the sun and its temperature increases. Due to different amount of heat energy receipt at different places, the air temperature at different places also vary. The variation in air temperature basically results into air motion, so as to equalize the energy content of the different regions of the earth. Thus temperature of air can be regarded as the basic cause for weather changes.

Qualification of Atmospheric Temperature

Atmospheric temperature is continuously changing; it is never steady or constant for a long time. Therefore quantification of atmospheric temperature is very important aspect. The atmospheric temperature can be quantified in the following ways.

Maximum temperature: It is the highest temperature attained by the atmosphere in diurnal variation.

Minimum temperature: It is the lowest temperature attained by the at by the atmosphere in diurnal variation.

Average temperature: It represents the average temperature condition of the atmosphere during 24 hours of the day.

Atmospheric Temperature Range

Atmospheric temperature range is the numerical difference between the minimum and maximum values of temperature observed in a given location during a period of time (e.g., in a given day, month, year, century) or the average (average of all temperature ranges in a period of time).

The variation in temperature that occurs from the highs of the day to the cool of nights is called diurnal temperature variation.

Parametres Affecting the Range

The size of ground-level atmospheric temperature ranges depends on several factors, such as:

- The average temperature
- The average humidity
- The regime of winds (intensity, duration, variation, temperature, etc.)
- The proximity to large bodies of water, such as the sea

The figure below shows an example of monthly temperatures recorded at one of such locations, the city of Campinas, state of São Paulo, Brazil, which lies approximately 60 km north of the Capricorn line (latitude of 22 degrees). Average yearly temperature is 22.4 degrees Celsius, ranging from an average minimum of 12.2 degrees to a maximum of 29.9 degrees.

The average temperature range is 11.4 degrees. Variability along the year is small (standard deviation of 2.31 for the maximum monthly average and 4.11 for the minimum). It is easy to see in the graph another typical phenomenon of temperature ranges, which is its increase during winter (lower average air temperature).

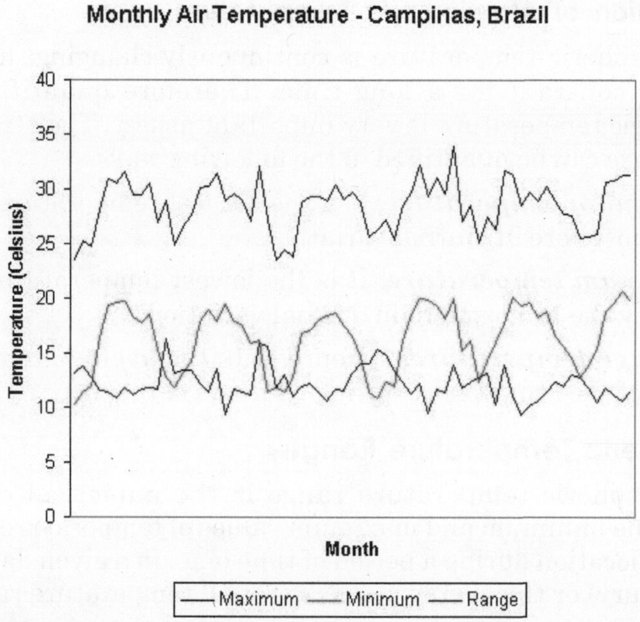

Figure: *Average maximum, minimum and range of monthly air temperatures recorded in Campinas, Brazil, between January 2001 and July 2006*

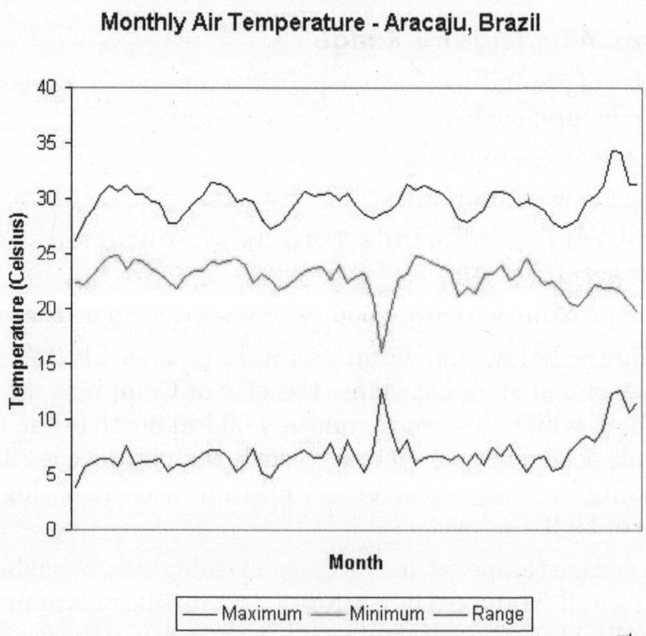

Figure: *Average maximum, minimum and range of monthly air temperatures recorded in Aracaju, state of Sergipe, Brazil, between January 2001 and July 2006*

In Campinas, for example, the daily temperature range in July (the coolest month of the year) may vary between typically 10 and 24 degrees Celsius (range of 14), while in January, it may range between 20 and 30 degrees Celsius (range of 10).

The effect of latitude, tropical climate, constant gentle wind and sea-side locations show smaller average temperature ranges, smaller variations of temperature, and a higher average temperature (second graph, taken for the same period as Campinas, at Aracaju, capital of the state of Sergipe, also in Brazil, at a latitude of 10 degrees, nearer to the Equator). Average maximum yearly temperature is 28.7 degrees Celsius and average minimum is 21.9. The average temperature range is 5.7 degrees only. Temperature variation along the year in Aracaju is very damped (standard deviation of 1.93 for the maximum temperature and 2.72 for the minimum temperature).

Uses

A location which combines an average temperature of 19 degrees Celsius, 60% average humidity and a temperature range of about 10 degrees Celsius around the average temperature (yearly temperature variation) is considered ideal in terms of comfort for the human species. Most of the places with these characteristics are located in the transition between temperate and tropical climates, approximately around the tropics, particularly in the Southern hemisphere (the tropic of Capricorn).

Lifted Minimum Temperature

The minimum temperature at night does not occur on the ground but few tens of centimetres above the ground. The lowest temperature layer is called *Ramdas layer* after L. A. Ramdas, who first reported this phenomenon in 1932 based on observations at different screen heights at six meteorological centres across India. The phenomenon is attributed to the interaction of thermal radiation effects on atmospheric aerosols and convection transfer close to the ground.

Factors Affecting the Air Temperature

The distribution of temperature over the earth surface depends on following factors:

Latitude: Highest temperatures are generally at the equator and the lowest at the poles. The temperature decreases with the increase of latitude.

Altitude: Temperature decreases with height in troposphere.

Season: Coldest temperatures are in winter and highest temperatures are in summer seasons.

Distribution of land and water: Water bodies are great moderators of temperature. Because of high Specific heat of water, so on the oceans, the regularity in temperature is more as compared to continents.

Topogrtaphy: Mountain ranges affect the temperature by acting as obstacles to the Flow of cold air cold air near the surface and they often set conditions of warm winds.

Ocean currents: Hot and cold ocean currents affect temperature e.g. Gulag Stream (Warm) in North Atlantic, Benguela current (cold) along West coast of South Africa, Peru Current (cold) along West Coast of South America.

Winds: Various types of wind affect temperature.

Clouds and rains: Clouds by obstructing the heat from the Sun and rains by cooling the Atmosphere, affect the temperature.

Colour of the soil: Black colour of soils absorbs more radiations and other types reflect them.

Slope of the soil: Black colour of soils absorbs more radiations and other types reflect them.

Forest and vegetation: Due to Evapotranspiration and interception of sun – rays, temperatures are moderated.

Soil Temperature

The temperature of a soil is important as it affects how fast plants can grow. Soil temperature also affects how quickly plants take up water and nutrients.

Clay soils are cold, wet soils. Germination and seedling growth is usually slow.

Because sandy soils don't contain much water but lots of air, they warm up quickly. They are useful for growing early crops.

Soil temperature affects the speed of chemical reactions. Warm temperatures speed up reactions and colder ones slow them down. Soil temperature affects the breakdown of parent material and how fast micro-organisms work.

Both are important in adding and returning nutrients to the soil. Soil temperature is influenced by the climate of the area and the season of the year.

Climate and Season

In a warm climate or during summer, the soil is full of chemical and physical activity.

Micro-organisms will slow down when it is too hot (above 35°C). Micro-organism also slows down in the middle of winter as the soil is too cold. They become inactive below 6°C.

In winter it takes a long time for organic matter to break down in soils in the cooler parts of New Zealand.

Plants are also affected by soil temperature. When the temperature is right for a particular plant its roots will keep growing. If the soil gets too hot or too cold then the uptake of water and the growth of roots will slow up.

Aspect

The slope of the land and the direction that it faces directly affects the temperature of a soil. Sun will fall on north-facing land during the day in both summer and winter.

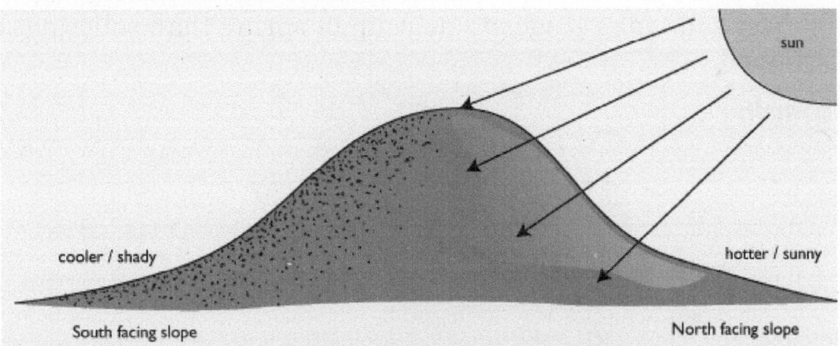

The effect of aspect on soil temperature

During the winter south-facing slopes will get less sun during the day. The soil on these slopes cools down quickly in the autumn and warms up slowly in the spring.

In the middle of summer these areas will have better growth because the soil is not as hot and dry as the soil on a north-facing slope.

Plant Cover

Soil covered in plants is protected (insulated) from fast heating or cooling of the soil (temperature fluctuations).

In a recently cultivated paddock, soil will heat up quickly during the day, but it will also lose heat quickly once the sun goes down.

Figure: *Cultivated soil*

Soil Colour

A dark coloured soil with a lot of organic matter in it heats quickly as it absorbs more heat energy.

Soil Water

Wet soil will be cooler because it takes a lot of energy to heat water. Wet soils take longer to heat up in spring than soils that are well drained.

Soil Depth

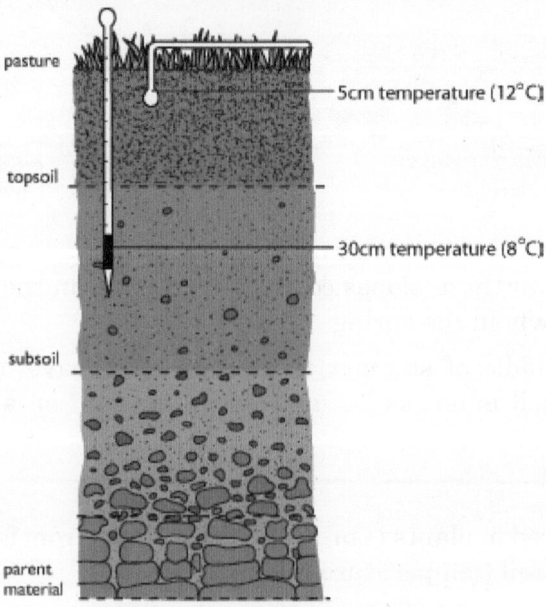

A soil profile showing thermometers at different depths on a summers day.

The deeper you go down in a soil profile the less the soil temperature will fluctuate. Soil is a good insulator. It can take a while for the soil at the bottom of a profile to heat up, but it will also take a longer time for it to lose the heat that is stored there.

Variations of Soil Temperature

There are two types of Soil Temperature:

1. Daily and Seasonal Variation of Soil Temperature.
   a) There variations occur at the surface of the soil.
   b) At 5 cm depth the change exceeds 10 0C At 20 cm the change is less and at 80 cm diurnal changes are practically nil
   c) On cooler days the changes are smaller due to increased best capacity as the soils become wetter on these days.
   d) On a clear sunny day a bare soil surface is hotter than the air temperature.
   e) The time of the peak temperature of the soil reaches earlier than the air temperature due to the lag of the air temperature.
   f) At around 20 cm in the soil the temperature in the ground reaches peak after the surface reaches its maximum due to more tune the heat takes to penetrate the soil. The rate of penetration of heat wave within the soil takes around 3 hours to reach 10 cm depth.
   g) The cooling period of the daily cycle of the soil surface temperature is almost double than the warning period.
   h) Undesirable daily temperature variations can be minimized by scheduling irrigation.

Seasonal Variations of Soil Temperature

a) Seasonal variations occur much deeper into the soil.
b) When the plant canopy is fully developed the seasonal variations are smaller.
c) In winter, the depth to which the soil freezes depends on the duration and severances of the winter.
d) In summer the soil temperature variations are much more than winter in trophies and sub trophies.

Soil Thermal Properties

The thermal properties of soil are a component of soil physics that has found important uses in engineering, climatology and agriculture.

These properties influence how energy is partitioned in the soil profile. While related to soil temperature, it is more accurately associated with the transfer of heat throughout the soil, by radiation, conduction and convection.

The main soil thermal properties are:

- Volumetric heat capacity, SI Units: $J.m^{-3}.K^{-1}$
- Thermal conductivity, SI Units: $W.m^{-1}.K^{-1}$
- Thermal diffusivity, SI Units: $m^2.s^{-1}$

Measurement

It is hard to say something general about the soil thermal properties at a certain location because these are in a constant state of flux from diurnal and seasonal variations. Apart from the basic soil composition, which is constant at one location, soil thermal properties are strongly influenced by the soil volumetric water content, volume fraction of solids and volume fraction of air. Air is a poor thermal conductor and reduces the effectiveness of the solid and liquid phases to conduct heat. While the solid phase has the highest conductivity it is the variability of soil moisture that largely determines thermal conductivity.

As such soil moisture properties and soil thermal properties are very closely linked and are often measured and reported together. Temperature variations are most extreme at the surface of the soil and these variations are transferred to sub surface layers but at reduced rates as depth increases. Additionally there is a time delay as to when maximum and minimum temperatures are achieved at increasing soil depth (sometimes referred to as thermal lag).

One possible way of assessing soil thermal properties is the analysis of soil temperature variations versus depth Fourier's law,

$$Q = -\lambda dT / dz$$

where Q is heat flux or rate of heat transfer per unit area $J \cdot m^{-2'} - s^{-1}$ or $W \cdot m^{-2}$, λ is thermal conductivity $W \cdot m^{-1}.K^{-1}$; dT/dz is the gradient of temperature (change in temp/change in depth) $K \cdot m^{-1}$.

The most commonly applied method for measurement of soil thermal properties, is to perform in-situ measurements, using Non-Steady-State Probe systems, or Heat Probes.

Single and Dual Heat Probes

The single probe method employs a heat source inserted into the soil whereby heat energy is applied continuously at a given rate. The thermal properties of the soil can be determined by analysing the temperature response adjacent to the heat source via a thermal sensor. This method reflects the rate at which heat is conducted away from the probe. The limitation of this device is that it measures thermal conductivity only. Applicable standards are: IEEE Guide for Soil Thermal Resistivity Measurements (IEEE Standard 442-1981) as well as with ASTM D 5334-08 Standard Test Method for Determination of Thermal Conductivity of Soil and Soft Rock by Thermal Needle Probe Procedure.

After further research the dual-probe heat-pulse technique was developed. It consists of two parallel needle probes separated by a distance (r). One probe contains a heater and the other a temperature sensor. The dual probe device is inserted into the soil and a heat pulse is applied and the temperature sensor records the response as a function of time. That is, a heat pulse is sent from the probe across the soil (r) to the sensor.

The great benefit of this device is that it measures both thermal diffusivity and volumetric heat capacity. From this, thermal conductivity can be calculated meaning the dual probe can determine all the main soil thermal properties. Potential drawbacks of the heat-pulse technique have been noted. This includes the small measuring volume of soil as well as measurements being sensitive to probe-to-soil contact and sensor-to-heater spacing.

Remote Sensing

Remote sensing from satellites, aircraft has greatly enhanced how the variation in soil thermal properties can be identified and utilized to benefit many aspects of human endeavour. While remote sensing of reflected light from surfaces does indicate thermal response of the topmost layers of soil (a few molecular layers thick), it is thermal infrared wavelength that provides energy variations extending to varying shallow depths below the ground surface which is of most interest. A thermal sensor can detect variations to heat transfers into and out of near surface layers because of external heating by the thermal processes of conduction, convection, and radiation. Microwave remote sensing from satellites has also proven useful as it has an advantage over TIR of not being effected by cloud cover.

The various methods of measuring soil thermal properties have been utilized to assist in diverse fields such as; the expansion and contraction of construction materials especially in freezing soils, longevity and efficiency of gas pipes or electrical cables buried in the ground, energy conservation schemes, in agriculture for timing of planting to ensure optimum seedling emergence and crop growth, measuring greenhouse gas emissions as heat effects the liberation of carbon dioxide from soil.

Soil thermal properties are also becoming important in areas of environmental science such as determining water movement in radioactive waste and in locating buried land mines.

Daily and Seasonal Patterns of Soil Temperature

The range of soil temperature in summer and winter decreases with increasing depth. At 40 cm depth the change is very minor and at 81 cm no diurnal change occurs.

In summer soil temperature decreases with depth during the day time.Temperature gradients direct heat into the soil. At night hours the temperature is highest between 20 and 40 cm, and from that level heat is directed both upward and downward.

In winter the 81 cm depth is warmer and the diurnal change is still very minor at 40 cm depth. Heat is transferred upward from those levels though out the day and night, because of the low radiation intensity in winter. The daily range of surface temperature is very small; however the range is still greater at the surface than any below level. Only during the hours near noon does heat penetrate into the soil from the surface.

Soil Temperature Profiles

The heat is continuously moving in to or out of the soil and thermal energy is being continuously redistributed in the soil. Heat will not flow under isothermal condition.

The pattern of soil temperature profiles changes rapidly during & normal day. The soil surface is the coldest level in the early morning and the warmest in the early afternoon.

At midday heat is directed downward through the upper one metro soil. The profiles show that the heat exists from middle of the layer after sunset but some heat flow continuously downward throughout the night. During most of the day temperature profile indicate downward heat flux. This is summer there is a net daily gain or storage of heat in the soil.

Factors Affecting the Soil Temperature and its Control

1. *Solar radiation:* The amount of heat from the Sun that reaches the earth is 2.0 cal/cm2 min -1 the amount of radiation received by the soil depends on angles with which the soil faces the Sun.

2. *Condensation:* Whenever water vapour from soil depths or atmosphere condenses in the soil, its heat increases noticeably.
3. *Evaporation:* The greater the rate of evaporation, the more the soil is cooled.
4. *Rainfall:* Rainfall cools down the soil.
5. *Vegetation:* A bare soil quickly absorbs heat and becomes very hot during the summer and become very cold during the winter. Vegetation acts as a insulating agent. It does not allow the soil to become either too hot during the summer and two cold during the winter.
6. *Colour of the soil:* Black coloured soils absorbs more heat than light closured soils Hence black colour soils are warmer than light coloured soils.
7. *Moisture content:* A soil with higher moisture content is cooler than dry soil.
8. *Tillage:* The cultivated soil has greater temperature amplitude as compared to the uncultivated soil.
9. *Soil texture:* Soil textures affect the thermal conductivity of soil. Thermal conductivity decreases with reduction in particle size.
10. *Organic matter content:* Organic matter reduces the heat capacity and thermal conductivity of soil, increases its water holding capacity and has a dark colour, which increases its heat absorbability.
11. *Slope of land:* Solar radiation that reaches the land surface at an angle is scattered over a wider area than the same amount of solar radiation reaching the surface of the land at right angles. Therefore, the amount of solar radiation reaching per unit area of the land surface decreases as the slope of the land is increases.

Soil Temperature can be Controlled by

1. Regulating soil moisture.
2. Proper soil management practices so a to have good drainage.
3. Application /use of mulching.
4. Sufficient addition of organic matter.

Air Pressure

Definition: Atmospheric pressure can be defined as the weight exerted by air column on units surface of the earth.

Units of pressure:

1. Height of mercury column measured in inches, cm, and mm
2. Bar, bar is a force equal to 106 dynes/cm2
3. SI (standard International) unit or pressure is Pascal

1. Pascal = force of 1 Newton/m2

= 1 N m2

Standard Atmospheric Pressure

The standard atmospheric pressure is given at mean sea level at 450K latitude and at temperature of 2730K

Standard Atmospheric = 29.92 inches or 76 cm or 760 mm

Pressure = 1013.25 mb

= 101.325 kilo Pascal (Kpa)

= 14.7 lbs/inch2

1.014 x 106 dynes/cm2

Isobar

Any line joining the places of equal atmospheric pressure on the weather map is known as isobar. Where isobars are closely spaced, a rapid of steep change in reassure is indicated. When isobars are widely speeding, a slow change in pressure is indicated. Two isobars are never cross each other. Isobars are plotted on the map to show the distribution of pressure. The isobars are drawn at pressure intervals of 2, 3,4 or 5 mb.

Pressure gradient: The rate of change of atmospheric pressure per unit horizontal distance between two points at the same elevation is known as pressure gradient or isobaric slope. This change tak3es place and is measured in the direction perpendicular to the isobars preferably from high to low pressures. This exerts a force on air particles and is important in determining the strength of wind. The pressure gradient is expressed in decrease in pressure per unit horizontal distance as mb/100 metres.

Variation in Atmospheric Pressure

Variation with height or vertical variation: The pressure depends on the density or mass of the air. The density of air depends on its temperature. Its composition and force of gravity. I t is observed that the density of air decreases with increase in height so the pressure also decreases with increase in eight.

The pressure at sea level is 1013.25 mb at 50 km height it becomes 0.93 mb and 80 km it is only 0.03 mb. This indicates how rapidly the atmospheric gas becomes thinner to decrease density and so the pressure. The pressure decreases on an average at the rate of about 34 mb per every 300 metres height.

Horizontal variation of pressure: The horizontal variation of atmospheric pressure depends on temperature, extent of water vapour, latitude and land and water relationship.

The equatorial low pressure belt : Along the equator lies a belt of low pressure known as the equatorial low or doldrums or calm. This low pressure belt lies between 50 North and 50 South latitudes.

Sub – tropical high pressure belt: The high pressure belt are found between 24 – 300C latitudes in both the hemispheres.

Low pressure belts near 600 latitudes: The airs from this area get thrown outwards on account of the rotation of the earth and this is how the low pressure belts are created.

Polar high pressure belts: The temperature is extremely low in the Polar Regions. The air being cold and heavy throughout the year a high pressure belt is created in both Polar Regions.

Diurnal variation: At a given station the pressure show the two high and two lows. On normal pressure day two maxima i.e. one at 10 a.m. and another at 10 p.m. and two minimasi.e. one at 4 a.m. and another at 4 p.m. are observed.

Thus there is double oscillation caused by alternate heating and cooling of atmosphere.

Factors Affecting Atmospheric Pressure

1. Temperature of air
2. Altitude
3. Water vapour in air
4. Revolution and gravitation of the earth.

Wind

Wind is the flow of gases on a large scale. On the surface of the Earth, wind consists of the bulk movement of air. In outer space, solar wind is the movement of gases or charged particles from the sun through space, while planetary wind is the outgassing of light chemical elements from a planet's atmosphere into space. Winds are commonly classified by their spatial scale, their speed, the types of forces that cause them, the regions in which they occur, and their effect. The strongest observed winds on a planet in our solar system occur on Neptune and Saturn.

Winds have various aspects, an important one being its velocity; another the density of the gas involved; another is the energy content or wind energy of a wind.

In meteorology, winds are often referred to according to their strength, and the direction from which the wind is blowing. Short bursts of high speed wind are termed gusts. Strong winds of intermediate duration (around one minute) are termed squalls. Long-duration winds have various names associated with their average strength, such as breeze, gale, storm, hurricane, and typhoon. Wind occurs on a range of scales, from thunderstorm flows lasting tens of minutes, to local breezes generated by heating of land surfaces and lasting a few hours, to global winds resulting from the difference in absorption of solar energy between the climate zones on Earth.

The two main causes of large-scale atmospheric circulation are the differential heating between the equator and the poles, and the rotation of the planet (Coriolis effect). Within the tropics, thermal low circulations over terrain and high plateaus can drive monsoon circulations. In coastal areas the sea breeze/land breeze cycle can define local winds; in areas that have variable terrain, mountain and valley breezes can dominate local winds.

In human civilization, wind has inspired mythology, influenced the events of history, expanded the range of transport and warfare, and provided a power source for mechanical work, electricity and recreation. Wind powers the voyages of sailing ships across Earth's oceans. Hot air balloons use the wind to take short trips, and powered flight uses it to increase lift and reduce fuel consumption. Areas of wind shear caused by various weather phenomena can lead to dangerous situations for aircraft. When winds become strong, trees and man-made structures are damaged or destroyed.

Winds can shape landforms, via a variety of aeolian processes such as the formation of fertile soils, such as loess, and by erosion. Dust from large deserts can be moved great distances from its source region by the prevailing winds; winds that are accelerated by rough topography and associated with dust outbreaks have been assigned regional names in various parts of the world because of their significant effects on those regions. Wind affects the spread of wildfires. Winds disperse seeds from various plants, enabling the survival and dispersal of those plant species, as well as flying insect populations. When combined with cold temperatures, wind has a negative impact on livestock. Wind affects animals' food stores, as well as their hunting and defensive strategies.

Cause

Wind is caused by differences in the atmospheric pressure. When a difference in atmospheric pressure exists, air moves from the higher to the lower pressure area, resulting in winds of various speeds. On a rotating planet, air will also be deflected by the Coriolis effect, except exactly on the equator. Globally, the two major driving factors of large-scale wind patterns (the atmospheric circulation) are the differential heating between the equator and the poles (difference in absorption of solar energy leading to buoyancy forces) and the rotation of the planet. Outside the tropics and aloft from frictional effects of the surface, the large-scale winds tend to approach geostrophic balance.

Near the Earth's surface, friction causes the wind to be slower than it would be otherwise. Surface friction also causes winds to blow more inward into low pressure areas. A new, controversial theory, suggests atmospheric gradients are caused by forest induced water condensation resulting in a positive feedback cycle of forests drawing moist air from the coastline.

Winds defined by an equilibrium of physical forces are used in the decomposition and analysis of wind profiles. They are useful for simplifying the atmospheric equations of motion and for making qualitative arguments about the horizontal and vertical distribution of winds. The geostrophic wind component is the result of the balance between Coriolis force and pressure gradient force. It flows parallel to isobars and approximates the flow above the atmospheric boundary layer in the midlatitudes. The thermal wind is the *difference* in the geostrophic wind between two levels in the atmosphere. It exists only in an atmosphere with horizontal temperature gradients. The ageostrophic wind component is the difference between actual and geostrophic wind, which is responsible for air "filling up" cyclones over time. The gradient wind is similar to the geostrophic wind but also includes centrifugal force (or centripetal acceleration).

Measurement

Wind direction is usually expressed in terms of the direction from which it originates. For example, a *northerly* wind blows from the north to the south. Weather vanes pivot to indicate the direction of the wind. At airports, windsocks indicate wind direction, and can also be used to estimate wind speed by the angle of hang. Wind speed is measured by anemometres, most commonly using rotating cups or propellers. When a high measurement frequency is needed (such as in research applications), wind can be measured by the propagation speed of ultrasound signals or by the effect of ventilation on the resistance of a heated wire. Another type of anemometre uses pitot tubes that take advantage of the pressure differential between an inner tube and an outer tube that is exposed to the wind to determine the dynamic pressure, which is then used to compute the wind speed.

Sustained wind speeds are reported globally at a 10 metres (33 ft) height and are averaged over a 10 minute time frame. The United States reports winds over a 1 minute average for tropical cyclones, and a 2 minute average within weather observations. India typically reports winds over a 3 minute average. Knowing the wind sampling average is important, as the value of a one-minute sustained wind is typically 14% greater than a ten-minute sustained wind. A short burst of high speed wind is termed a wind gust, one technical definition of a wind gust is: the maxima that exceed the lowest wind speed measured during a ten-minute time interval by 10 knots (19 km/h). A squall is a doubling of the wind speed above a certain threshold, which lasts for a minute or more.

To determine winds aloft, rawinsondes determine wind speed by GPS, radio navigation, or radar tracking of the probe. Alternatively, movement of the parent weather balloon position can be tracked from the ground visually using theodolites. Remote sensing techniques for wind include SODAR, Doppler lidars and radars, which can measure the Doppler shift of electromagnetic radiation scattered or reflected off suspended aerosols or molecules, and radiometres and radars can be used to measure the surface roughness of the ocean from space or airplanes. Ocean roughness can be used to estimate wind velocity close to the sea surface over oceans. Geostationary satellite imagery can be used to estimate the winds throughout the atmosphere based upon how far clouds move from one image to the next. Wind Engineering describes the study of the effects of the wind on the built environment, including buildings, bridges and other man-made objects.

Wind Force Scale

Historically, the Beaufort wind force scale provides an empirical description of wind speed based on observed sea conditions. Originally it was a 13-level scale, but during the 1940s, the scale was expanded to 17 levels. There are general terms that differentiate winds of different average speeds such as a breeze, a gale, a storm, tornado, or a hurricane. Within the Beaufort scale, gale-force winds lie between 28 knots (52 km/h) and 55 knots (102 km/h) with preceding adjectives such as moderate, fresh, strong, and whole used to differentiate the wind's strength within the gale category. A storm has winds of 56 knots (104 km/h) to 63 knots (117 km/h). The terminology for tropical cyclones differs from one region to another globally. Most ocean basins use the average wind speed to determine the tropical cyclone's category. Below is a summary of the classifications used by Regional Specialized Meteorological Centres worldwide:

General wind classifications			*Tropical cyclone classifications (all winds are 10-minute averages)*					
Beaufort scale	***10-minute sustained winds (knots)***	***General term***	***N Indian Ocean IMD***	***SW Indian Ocean MF***	***Australian region South Pacific BoM, BMKG, FMS, MSNZ***	***NW Pacific JMA***	***NW Pacific JTWC***	***NE Pacific & N Atlantic NHC & CPHC***
0	<1	Calm	Low Pressure Area	Tropical disturbance	Tropical low Tropical Depression	Tropical depression	Tropical depression	Tropical depression
1	1–	Light air						
2	4–6	Light breeze						
3	7–10	Gentle breeze						
4	11–16	Moderate breeze						
5	17–21	Fresh breeze	Depression					
6	22–27	Strong breeze						
7	28–29	Moderate gale	Deep depression	Tropical depression				
	30–33							
8	34–40	Fresh gale	Cyclonic storm	Moderate tropical storm	Tropical cyclone (1)	Tropical storm	Tropical storm	Tropical storm
9	41–47	Strong gale						
10	48–55	Whole gale	Severe cyclonic storm	Severe tropical storm	Tropical cyclone (2)	Severe tropical storm		
11	56–63	Storm						
12	64–72	Hurricane	Very severe cyclonic storm	Tropical cyclone	Severe tropical cyclone (3)	Typhoon	Typhoon	Hurricane (1)
13	73–85							Hurricane (2)
14	86–89				Severe tropical cyclone (4)			Major hurricane (3)
15	90–99			Intense tropical cyclone				
16	100–106							Major hurricane (4)
17	107–114				Severe tropical cyclone (5)			
	115–119			Very intense tropical cyclone			Super typhoon	
	>120		Super cyclonic storm					Major hurricane (5)

Station Model

The station model plotted on surface weather maps uses a wind barb to show both wind direction and speed. The wind barb shows the speed using "flags" on the end.

- Each half of a flag depicts 5 knots (9.3 km/h) of wind.
- Each full flag depicts 10 knots (19 km/h) of wind.
- Each pennant (filled triangle) depicts 50 knots (93 km/h) of wind.

Winds are depicted as blowing from the direction the barb is facing. Therefore, a northeast wind will be depicted with a line extending from the cloud circle to the northeast, with flags indicating wind speed on the northeast end of this line.

Once plotted on a map, an analysis of isotachs (lines of equal wind speeds) can be accomplished. Isotachs are particularly useful in diagnosing the location of the jet stream on upper level constant pressure charts, and are usually located at or above the 300 hPa level.

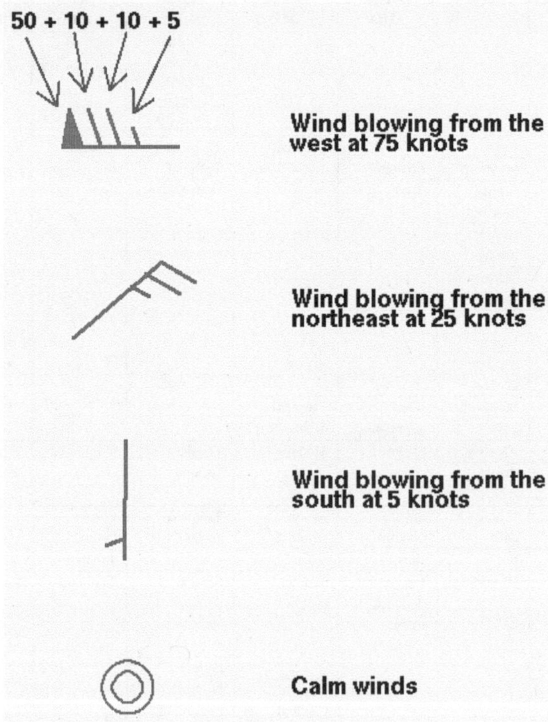

Figure: *Wind plotting within a station model*

Wind Power

Wind energy is the kinetic energy of the air in motion. The kinetic energy of a packet of air of mass m with velocity v is given by $\frac{1}{2} m v^2$.

To find the mass of the packet passing through an area A perpendicular its velocity (which could be the rotor area of a turbine), we multiply its volume after time t has passed with the air density ρ, which gives us $m = A\, v\, t\, \rho$. So, we find that the total wind energy is:

$$E = \frac{1}{2}\rho A v^3 t$$

Differentiating with respect to time to find the rate of increase of energy, we find that the total wind power is:

$$P = dE / dt = \frac{1}{2}\rho A v^3$$

Wind power is thus proportional to the third power of the wind velocity.

Theoretical Power Captured by a Wind Turbine

Total wind power could be captured only if the wind velocity is reduced to zero. In a realistic wind turbine this is impossible, as the captured air must also leave the turbine. A relation between the input and output wind velocity must be considered. Using the concept of stream tube, the maximal achievable extraction of wind power by a wind turbine is 59% of the total theoretical wind power.

Practical Wind Turbine Power

Further insufficiencies, such as rotor blade friction and drag, gearbox losses, generator and converter losses, reduce the power delivered by a wind turbine. The basic relation that the turbine power is (approximately) proportional to the third power of velocity remains.

Global Climatology

Easterly winds, on average, dominate the flow pattern across the poles, westerly winds blow across the mid-latitudes of the earth, polewards of the subtropical ridge, while easterlies again dominate the tropics.

Directly under the subtropical ridge are the doldrums, or horse latitudes, where winds are lighter. Many of the Earth's deserts lie near the average latitude of the subtropical ridge, where descent reduces the relative humidity of the air mass. The strongest winds are in the mid-latitudes where cold polar air meets warm air from the tropics.

Tropics

The trade winds (also called trades) are the prevailing pattern of easterly surface winds found in the tropics towards the Earth's equator. The trade winds blow predominantly from the northeast in the Northern Hemisphere and from the southeast in the Southern Hemisphere. The trade winds act as the steering flow for tropical cyclones that form over world's oceans. Trade winds also steer African dust westward across the Atlantic Ocean into the Caribbean Sea, as well as portions of southeast North America.

A monsoon is a seasonal prevailing wind that lasts for several months within tropical regions. The term was first used in English in India, Bangladesh, Pakistan, and neighbouring countries to refer to the big seasonal winds blowing from the Indian Ocean and Arabian Sea in the southwest bringing heavy rainfall to the area. Its poleward progression is accelerated by the development off a heat low over the Asian, African, and North American continents during May through July, and over Australia in December.

Westerlies and their Impact

The Westerlies or the Prevailing Westerlies are the prevailing winds in the middle latitudes between 35 and 65 degrees latitude. These prevailing winds blow from the west to the east, and steer extratropical cyclones in this general manner. The winds are predominantly from the southwest in the Northern Hemisphere and from the northwest in the Southern Hemisphere. They are strongest in the winter when the pressure is lower over the poles, and weakest during the summer and when pressures are higher over the poles.

Together with the trade winds, the westerlies enabled a round-trip trade route for sailing ships crossing the Atlantic and Pacific Oceans, as the westerlies lead to the development of strong ocean currents on the western sides of oceans in both hemispheres through the process of western intensification.

These western ocean currents transport warm, sub tropical water polewards toward the polar regions. The westerlies can be particularly strong, especially in the southern hemisphere, where there is less land in the middle latitudes to cause the flow pattern to amplify, which slows the winds down. The strongest westerly winds in the middle latitudes are within a band known as the Roaring Forties, between 40 and 50 degrees latitude south of the equator.

The Westerlies play an important role in carrying the warm, equatorial waters and winds to the western coasts of continents, especially in the southern hemisphere because of its vast oceanic expanse.

Polar Easterlies

The polar easterlies, also known as Polar Hadley cells, are dry, cold prevailing winds that blow from the high-pressure areas of the polar highs at the north and south poles towards the low-pressure areas within the Westerlies at high latitudes. Unlike the Westerlies, these prevailing winds blow from the east to the west, and are often weak and irregular. Because of the low sun angle, cold air builds up and subsides at the pole creating surface high-pressure areas, forcing an equatorward outflow of air; that outflow is deflected westward by the Coriolis effect.

Local Considerations

In coastal regions, sea breezes and land breezes can be important factors in a location's prevailing winds. The sea is warmed by the sun more slowly because of water's greater specific heat compared to land. As the temperature of the surface of the land rises, the land heats the air above it by conduction. The warm air is less dense than the surrounding environment and so it rises. This causes a pressure gradient of about 2 millibars from the ocean to the land. The cooler air above the sea, now with higher sea level pressure, flows inland into the lower pressure, creating a cooler breeze near the coast. When large-scale winds are calm, the strength of the sea breeze is directly proportional to the temperature difference between the land mass and the sea. If an offshore wind of 8 knots (15 km/h) exists, the sea breeze is not likely to develop.

At night, the land cools off more quickly than the ocean because of differences in their specific heat values. This temperature change causes the daytime sea breeze to dissipate. When the temperature onshore cools below the temperature offshore, the pressure over the water will be lower than that of the land, establishing a land breeze, as long as an onshore wind is not strong enough to oppose it.

Near Mountains

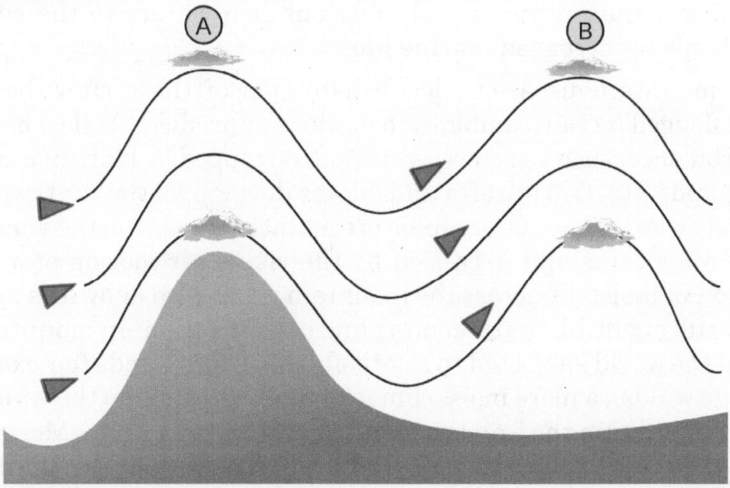

Figure: *Mountain wave schematic. The wind flows towards a mountain and produces a first oscillation (A). A second wave occurs further away and higher. The lenticular clouds form at the peak of the waves (B).*

Over elevated surfaces, heating of the ground exceeds the heating of the surrounding air at the same altitude above sea level, creating an associated thermal low over the terrain and enhancing any thermal lows that would have otherwise existed, and changing the wind circulation of the region. In areas where there is rugged topography that significantly interrupts the environmental wind flow, the wind circulation between mountains and valleys is the most important contributor to the prevailing winds. Hills and valleys substantially distort the airflow by increasing friction between the atmosphere and landmass by acting as a physical block to the flow, deflecting the wind parallel to the range just upstream of the topography, which is known as a barrier jet. This barrier jet can increase the low level wind by 45%. Wind direction also changes because of the contour of the land.

If there is a pass in the mountain range, winds will rush through the pass with considerable speed because of the Bernoulli principle that describes an inverse relationship between speed and pressure. The airflow can remain turbulent and erratic for some distance downwind into the flatter countryside. These conditions are dangerous to ascending and descending airplanes.

Cool winds accelerating through mountain gaps have been given regional names. In Central America, examples include the Papagayo wind, the Panama wind, and the Tehuano wind. In Europe, similar winds are known as the Bora, Tramontane, and Mistral. When these winds blow over open waters, they increase mixing of the upper layers of the ocean that elevates cool, nutrient rich waters to the surface, which leads to increased marine life.

In mountainous areas, local distortion of the airflow becomes severe. Jagged terrain combines to produce unpredictable flow patterns and turbulence, such as rotors, which can be topped by lenticular clouds. Strong updrafts, downdrafts and eddies develop as the air flows over hills and down valleys. Orographic precipitation occurs on the windward side of mountains and is caused by the rising air motion of a large-scale flow of moist air across the mountain ridge, also known as upslope flow, resulting in adiabatic cooling and condensation. In mountainous parts of the world subjected to relatively consistent winds (for example, the trade winds), a more moist climate usually prevails on the windward side of a mountain than on the leeward or downwind side. Moisture is removed by orographic lift, leaving drier air on the descending and generally warming, leeward side where a rain shadow is observed. Winds that flow over mountains down into lower elevations are known as downslope winds.

These winds are warm and dry. In Europe downwind of the Alps, they are known as foehn. In Poland, an example is the halny wiatr. In Argentina, the local name for downsloped winds is zonda. In Java, the local name for such winds is koembang. In New Zealand, they are known as the Nor'west arch, and are accompanied by the cloud formation they are named after that has inspired artwork over the years. In the Great Plains of the United States, the winds are known as a chinook. In California, downsloped winds are funnelled through mountain passes, which intensify their effect, and cxamples into Santa Ana and sundowner winds. Wind speeds during downslope wind effect can exceed 160 kilometres per hour (99 mph).

Average Wind Speeds

As described earlier, prevailing and local winds are not spread evenly across the earth, which means that wind speeds also differ by region. In addition, the wind speed also increases with the altitude.

Wind Power Density

Nowadays, a yardstick used to determine the best locations for wind energy development is referred to as wind power density (WPD). It is a calculation relating to the effective force of the wind at a particular location, frequently expressed in terms of the elevation above ground level over a period of time. It takes into account wind velocity and mass. Colour coded maps are prepared for a particular area are described as, for example, "mean annual power density at 50 metres." The results of the above calculation are included in an index developed by the National Renewable Energy Lab and referred to as "NREL CLASS." The larger the WPD calculation, the higher it is rated by class. At the end of 2008, worldwide nameplate capacity of wind-powered generators was 120.8 gigawatts.

Although wind produces only about 1.5% of worldwide electricity use, it is growing rapidly, having doubled in the three years between 2005 and 2008. In several countries it has achieved relatively high levels of penetration, accounting for approximately 19% of electricity production in Denmark, 10% in Spain and Portugal, and 7% in Germany and the Republic of Ireland in 2008. One study indicates that an entirely renewable energy supply based on 70% wind is attainable at today's power prices by linking wind farms with an HVDC supergrid.

Shear

Wind shear, sometimes referred to as windshear or wind gradient, is a difference in wind speed and direction over a relatively short distance in the Earth's atmosphere. Wind shear can be broken down into vertical and horizontal components, with horizontal wind shear seen across weather fronts and near the coast, and vertical shear typically near the surface, though also at higher levels in the atmosphere near upper level jets and frontal zones aloft.

Wind shear itself is a microscale meteorological phenomenon occurring over a very small distance, but it can be associated with mesoscale or synoptic scale weather features such as squall lines and cold fronts. It is commonly observed near microbursts and downbursts caused by thunderstorms, weather fronts, areas of locally higher low level winds referred to as low level jets, near mountains, radiation inversions that occur because of clear skies and calm winds, buildings, wind turbines, and sailboats. Wind shear has a significant effect during take-off and landing of aircraft because of their effects on control of the aircraft, and was a significant cause of aircraft accidents involving large loss of life within the United States.

Sound movement through the atmosphere is affected by wind shear, which can bend the wave front, causing sounds to be heard where they normally would not, or vice versa. Strong vertical wind shear within the troposphere also inhibits tropical cyclone development, but helps to organize individual thunderstorms into living longer life cycles that can then produce severe weather. The thermal wind concept explains how differences in wind speed with height are dependent on horizontal temperature differences, and explains the existence of the jet stream.

Usage of Wind

History

As a natural force, the wind was often personified as one or more wind gods or as an expression of the supernatural in many cultures. Vayu is the Hindu God of Wind. The Greek wind gods include Boreas, Notus, Eurus, and Zephyrus. Aeolus, in varying interpretations the ruler or keeper of the four winds, has also been described as Astraeus, the god of dusk who fathered the four winds with Eos, goddess of dawn. The Ancient Greeks also observed the seasonal change of the winds, as evidenced by the Tower of the Winds in Athens. Venti are the Roman gods of the winds. Fûjin, is the Japanese wind god and is one of the eldest Shinto gods. According to legend, he was present at the creation of the world and first let the winds out of his bag to clear the world of mist. In Norse mythology, Njord is the god of the wind. There are also four dvärgar (Norse dwarves), named Norπri, Suðri, Austri and Vestri, and probably the four stags of Yggdrasil, personify the four winds, and parallel the four Greek wind gods. Stribog is the name of the Slavic god of winds, sky and air. He is said to be the ancestor (grandfather) of the winds of the eight directions.

Kamikaze is a Japanese word, usually translated as divine wind, believed to be a gift from the gods. The term is first known to have been used as the name of a pair or series of typhoons that are said to have saved Japan from two Mongol fleets under Kublai Khan that attacked Japan in 1274 and again in 1281. Protestant Wind is a name for the storm that deterred the Spanish Armada from an invasion of England in 1588 where the wind played a pivotal role, or the favourable winds that enabled William of Orange to invade England in 1688. During Napoleon's Egyptian Campaign, the French soldiers had a hard time with the khamsin wind: when the storm appeared "as a blood-stint in the distant sky", the natives went to take cover, while the French "did not react until it was too late, then choked and fainted in the blinding, suffocating walls of dust." During the North African Campaign of the World War II, "allied and German troops were several times forced to halt in mid-battle because of sandstorms caused by khamsin ... Grains of sand whirled by the wind blinded the soldiers and created electrical disturbances that rendered compasses useless."

Transportation

There are many different forms of sailing ships, but they all have certain basic things in common. Except for rotor ships using the Magnus effect, every sailing ship has a hull, rigging and at least one mast to hold up the sails that use the wind to power the ship. Ocean journeys by sailing ship can take many months, and a common hazard is becoming becalmed because of lack of wind, or being blown off course by severe storms or winds that do not allow progress in the desired direction. A severe storm could lead to shipwreck, and the loss of all hands. Sailing ships can only carry a certain quantity of supplies in their hold, so they have to plan long voyages carefully to include appropriate provisions, including fresh water.

For aerodynamic aircraft which operate relative to the air, winds affect groundspeed, and in the case of lighter-than-air vehicles, wind may play a significant or solitary role in their movement and ground track. The velocity of surface wind is generally the primary factor governing the direction of flight operations at an airport, and airfield runways are aligned to account for the common wind direction(s) of the local area. While taking off with a tailwind may be necessary under certain circumstances, a headwind is generally desirable. A tailwind increases takeoff distance required and decreases the climb gradient.

Power Source

Historically, the ancient Sinhalese of Anuradhapura and in other cities around Sri Lanka used the monsoon winds to power furnaces as early as 300 BCE. The furnaces were constructed on the path of the monsoon winds to exploit the wind power, to bring the temperatures inside up to 1,200 °C (2,190 °F). An early historical reference to a rudimentary windmill was used to power an organ in the first century CE. The first practical windmills were later built in Sistan, Afghanistan, from the 7th century CE. These were vertical-axle windmills, which had long vertical driveshafts with rectangle shaped blades.

Made of six to twelve sails covered in reed matting or cloth material, these windmills were used to grind corn and draw up water, and were used in the gristmilling and sugarcane industries. Horizontal-axle windmills were later used extensively in Northwestern Europe to grind flour beginning in the 1180s, and many Dutch windmills still exist. High altitude wind power is the focus of over 30 companies worldwide using tethered technology rather than ground-hugging compressive-towers. Oil is being saved by using wind for powering cargo ships by use of the mechanical energy converted from the wind's kinetic energy using very large kites.

Recreation

Wind figures prominently in several popular sports, including recreational hang gliding, hot air ballooning, kite flying, snowkiting, kite landboarding, kite surfing, paragliding, sailing, and windsurfing. In gliding, wind gradients just above the surface affect the takeoff and landing phases of flight of a glider. Wind gradient can have a noticeable effect on ground launches, also known as winch launches or wire launches. If the wind gradient is significant or sudden, or both, and the pilot maintains the same pitch attitude, the indicated airspeed will increase, possibly exceeding the maximum ground launch tow speed. The pilot must adjust the airspeed to deal with the effect of the gradient. When landing, wind shear is also a hazard, particularly when the winds are strong.

As the glider descends through the wind gradient on final approach to landing, airspeed decreases while sink rate increases, and there is insufficient time to accelerate prior to ground contact. The pilot must anticipate the wind gradient and use a higher approach speed to compensate for it.

Role in the natural world

In arid climates, the main source of erosion is wind. The general wind circulation moves small particulates such as dust across wide oceans thousands of kilometres downwind of their point of origin, which is known as deflation. Westerly winds in the mid-latitudes of the planet drive the movement of ocean currents from west to east across the world's oceans. Wind has a very important role in aiding plants and other immobile organisms in dispersal of seeds, spores, pollen, etc. Although wind is not the primary form of seed dispersal in plants, it provides dispersal for a large percentage of the biomass of land plants.

Erosion

Erosion can be the result of material movement by the wind. There are two main effects. First, wind causes small particles to be lifted and therefore moved to another region. This is called deflation. Second, these suspended particles may impact on solid objects causing erosion by abrasion (ecological succession). Wind erosion generally occurs in areas with little or no vegetation, often in areas where there is insufficient rainfall to support vegetation. An example is the formation of sand dunes, on a beach or in a desert. Loess is a homogeneous, typically nonstratified, porous, friable, slightly coherent, often calcareous, fine-grained, silty, pale yellow or buff, windblown (Aeolian) sediment. It generally occurs as a widespread blanket deposit that covers areas of hundreds of square kilometres and tens of metres thick. Loess often stands in either steep or vertical faces. Loess tends to develop into highly rich soils. Under appropriate climatic conditions, areas with loess are among the most agriculturally productive in the world. Loess deposits are geologically unstable by nature, and will erode very readily. Therefore, windbreaks (such as big trees and bushes) are often planted by farmers to reduce the wind erosion of loess.

Desert Dust Migration

During mid-summer (July), the westward-moving trade winds south of the northward-moving subtropical ridge expand northwestward from the Caribbean Sea into southeastern North America. When dust from the Sahara moving around the southern periphery of the ridge within the belt of trade winds moves over land, rainfall is suppressed and the sky changes from a blue to a white appearance, which leads to an increase in red sunsets. Its presence negatively impacts air quality by adding to the count of airborne particulates. Over 50% of the African dust that reaches the United States affects Florida. Since 1970, dust outbreaks have worsened because of periods of drought in Africa. There is a large variability in the dust transport to the Caribbean and Florida from year to year. Dust events have been linked to a decline in the health of coral reefs across the Caribbean and Florida, primarily since the 1970s. Similar dust plumes originate in the Gobi Desert, which combined with pollutants, spread large distances downwind, or eastward, into North America.

There are local names for winds associated with sand and dust storms. The Calima carries dust on southeast winds into the Canary islands. The Harmattan carries dust during the winter into the Gulf of Guinea. The Sirocco brings dust from north Africa into southern Europe because of the movement of extratropical cyclones through the Mediterranean Sea. Spring storm systems moving across the eastern Mediterranean Sea cause dust to carry across Egypt and the Arabian peninsula, which are locally known as Khamsin. The Shamal is caused by cold fronts lifting dust into the atmosphere for days at a time across the Persian Gulf states.

Effect on Plants

Wind dispersal of seeds, or anemochory, is one of the more primitive means of dispersal. Wind dispersal can take on one of two primary forms: seeds can float on the breeze or alternatively, they can flutter to the ground. The classic examples of these dispersal mechanisms include dandelions (*Taraxacum* spp., Asteraceae), which have a feathery pappus attached to their seeds and can be dispersed long distances, and maples (*Acer (genus)* spp., Sapindaceae), which have winged seeds and flutter to the ground. An important constraint on wind dispersal is the need for abundant seed production to maximize the likelihood of a seed landing in a site suitable for germination.

There are also strong evolutionary constraints on this dispersal mechanism. For instance, species in the Asteraceae on islands tended to have reduced dispersal capabilities (i.e., larger seed mass and smaller pappus) relative to the same species on the mainland. Reliance upon wind dispersal is common among many weedy or ruderal species. Unusual mechanisms of wind dispersal include tumbleweeds. A related process to anemochory is anemophily, which is the process where pollen is distributed by wind. Large families of plants are pollinated in this manner, which is favoured when individuals of the dominant plant species are spaced closely together.

Wind also limits tree growth. On coasts and isolated mountains, the tree line is often much lower than in corresponding altitudes inland and in larger, more complex mountain systems, because strong winds reduce tree growth. High winds scour away thin soils through erosion, as well as damage limbs and twigs. When high winds knock down or uproot trees, the process is known as windthrow. This is most likely on windward slopes of mountains, with severe cases generally occurring to tree stands that are 75 years or older. Plant varieties near the coast, such as the Sitka spruce and sea grape, are pruned back by wind and salt spray near the coastline.

Wind can also cause plants damage through sand abrasion. Strong winds will pick up loose sand and topsoil and hurl it through the air at speeds ranging from 25–40 miles per hour. Such windblown sand causes extensive damage to plant seedlings because it ruptures plant cells, making them vulnerable to evaporation and drought. Using a mechanical sandblaster in a laboratory setting, scientists affiliated with the Agricultural Research Service studied the effects of windblown sand abrasion on cotton seedlings. The study showed that the seedlings responded to the damage created by the windblown sand abrasion by shifting energy from stem and root growth to the growth and repair of the damaged stems. After a period of four weeks the growth of the seedling once again became uniform throughout the plant, as it was before the windblown sand abrasion occurred.

Effect on Animals

Cattle and sheep are prone to wind chill caused by a combination of wind and cold temperatures, when winds exceed 40 kilometres per hour (25 mph), rendering their hair and wool coverings ineffective. Although penguins use both a layer of fat and feathers to help guard against coldness in both water and air, their flippers and feet are less immune to the cold. In the coldest climates such as Antarctica, emperor penguins use huddling behaviour to survive the wind and cold, continuously alternating the members on the outside of the assembled group, which reduces heat loss by 50%. Flying insects, a subset of arthropods, are swept along by the prevailing winds, while birds follow their own course taking advantage of wind conditions, in order to either fly or glide. As such, fine line patterns within weather radar imagery, associated with converging winds, are dominated by insect returns. Bird migration, which tends to occur overnight within the lowest 7,000 feet (2,100 m) of the Earth's atmosphere, contaminates wind profiles gathered by weather radar, particularly the WSR-88D, by increasing the environmental wind returns by 15 knots (28 km/h) to 30 knots (56 km/h).

Pikas use a wall of pebbles to store dry plants and grasses for the winter in order to protect the food from being blown away. Cockroaches use slight winds that precede the attacks of potential predators, such as toads, to survive their encounters. Their cerci are very sensitive to the wind, and help them survive half of their attacks. Elk have a keen sense of smell that can detect potential upwind predators at a distance of 0.5 miles (800 m). Increases in wind above 15 kilometres per hour (9.3 mph) signals glaucous gulls to increase their foraging and aerial attacks on thick-billed murres.

Related Damage

High winds are known to cause damage, depending upon their strength. Infrequent wind gusts can cause poorly designed suspension bridges to sway. When wind gusts are at a similar frequency to the swaying of the bridge, the bridge can be destroyed more easily, such as what occurred with the Tacoma Narrows Bridge in 1940. Wind speeds as low as 23 knots (43 km/h) can lead to power outages due to tree branches disrupting the flow of energy through power lines. While no species of tree is guaranteed to stand up to hurricane-force winds, those with shallow roots are more prone to uproot, and brittle trees such as eucalyptus, sea hibiscus, and avocado are more prone to damage. Hurricane-force winds cause substantial damage to mobile homes, and begin to structurally damage homes with foundations. Winds of this strength due to downsloped winds off terrain have been known to shatter windows and sandblast paint from cars. Once winds exceed 135 knots (250 km/h), homes completely collapse, and significant damage is done to larger buildings. Total destruction to man-made structures occurs when winds reach 175 knots (324 km/h). The Saffir-Simpson scale and Enhanced Fujita scale were designed to help estimate wind speed from the damage caused by high winds related to tropical cyclones and tornadoes, and vice versa.

Australia's Barrow Island holds the record for the strongest wind gust, reaching 408 km/h (253 mph) during tropical cyclone Olivia on 10 April 1996, surpassing the previous record of 372 km/h (231 mph) set on Mount Washington (New Hampshire) on the afternoon of 12 April 1934. The most powerful gusts of wind on Earth were created by nuclear detonations. The blast wave is similar to a strong wind gust over the ground. The largest nuclear explosion (50–58 megatons at an altitude of about 13,000 ft) generated a 20 bar blast pressure at ground zero, which is similar to a wind gust of 3,100 miles per hour.

Wildfire intensity increases during daytime hours. For example, burn rates of smoldering logs are up to five times greater during the day because of lower humidity, increased temperatures, and increased wind speeds. Sunlight warms the ground during the day and causes air currents to travel uphill, and downhill during the night as the land cools. Wildfires are fanned by these winds and often follow the air currents over hills and through valleys. United States wildfire operations revolve around a 24-hour *fire day* that begins at 10:00 a.m. because of the predictable increase in intensity resulting from the daytime warmth.

In Outer Space

The solar wind is quite different from a terrestrial wind, in that its origin is the sun, and it is composed of charged particles that have escaped the sun's atmosphere. Similar to the solar wind, the planetary wind is composed of light gases that escape planetary atmospheres. Over long periods of time, the planetary wind can radically change the composition of planetary atmospheres.

Planetary Wind

The hydrodynamic wind within the upper portion of a planet's atmosphere allows light chemical elements such as hydrogen to move up to the exobase, the lower limit of the exosphere, where the gases can then reach escape velocity, entering outer space without impacting other particles of gas. This type of gas loss from a planet into space is known as planetary wind. Such a process over geologic time causes water-rich planets such as the Earth to evolve into planets like Venus. Additionally, planets with hotter lower atmospheres could accelerate the loss rate of hydrogen.

Solar Wind

Rather than air, the solar wind is a stream of charged particles—a plasma—ejected from the upper atmosphere of the sun at a rate of 400 kilometres per second (890,000 mph). It consists mostly of electrons and protons with energies of about 1 keV. The stream of particles varies in temperature and speed with the passage of time. These particles are able to escape the sun's gravity, in part because of the high temperature of the corona, but also because of high kinetic energy that particles gain through a process that is not well-understood. The solar wind creates the Heliosphere, a vast bubble in the interstellar medium surrounding the solar system. Planets require large magnetic fields in order to reduce the ionization of their upper atmosphere by the solar wind. Other phenomena caused by the solar wind include geomagnetic storms that can knock out power grids on Earth, the aurorae such as the Northern Lights, and the plasma tails of comets that always point away from the sun.

On Other Planets

Strong 300 kilometres per hour (190 mph) winds at Venus's cloud tops circle the planet every four to five earth days. When the poles of Mars are exposed to sunlight after their winter, the frozen CO_2 sublimates, creating significant winds that sweep off the poles as fast as 400 kilometres per hour (250 mph), which subsequently transports large amounts of dust and water vapour over its landscape. Other Martian winds have resulted in cleaning events and dust devils. On Jupiter, wind speeds of 100 metres per second (220 mph) are common in zonal jet streams. Saturn's winds are among the solar system's fastest. Cassini–Huygens data indicated peak easterly winds of 375 metres per second (840 mph). On Uranus, northern hemisphere wind speeds reach as high as 240 metres per second (540 mph) near 50 degrees north latitude. At the cloud tops of Neptune, prevailing winds range in speed from 400 metres per second (890 mph) along the equator to 250 metres per second (560 mph) at the poles. At 70° S latitude on Neptune, a high-speed jet stream travels at a speed of 300 metres per second (670 mph).

Chapter 10

Importance or Role or Effects of Wind in Agriculture

Wind Direction and Wind Speed

A wind is named according to the direction from which it blows e.g. a wind coming from west is called west wind.

The direction from which wind blows is termed as windward direction and that to which it blows is termed as leeward direction.

Classification of Wind

Vertical current	*Horizontal*		
	Periodic	Regular	Local
1. Divergence	1.Monsoon Winds	1. Planetary Earth's general circulation	1.Land and sea breeze
2. Convergence			2. Mountain and valley breeze
3. Eddies			3. Foehn/Chinnook winds
4. Convection			4. Tornadoes

Earth's General Circulation System (Surface Wind)

The earth's surface wind system or earth's general circulation of wind can be represented by a simple model. In this model, the earth surface been considered uniform, means either all and or all water and the effects of local systems have been ignored, therefore the actual wind system is much more complicated than the described in the model.

It is to be noted that unequal heating of the earth's surface generate pressure gradient which give rise to wind. There are three latitudinal circulations and there are also important longitudinal variations around each hemisphere.

Trade winds: The condition of greatest heating and expansion at the equator causes rising of air and creating low pressure belt (50 N to 50S latitude) known as doldrums or equatorial low or calm. The rising of air from equator causes increase in pressure at 350 N 350S which is known as sub-tropical high or Horse latitude belt. The winds therefore flow from horse latitude belt. The winds therefore flow from horse latitude to the equatorial region called "Trade winds" While moving these winds, they are deflected by corollas force to the right in the and nor them hemisphere and to left in southern hemisphere and become North – East trades and south –East Trades in northern and southern hemispheres respectively.

The blow of air from equator and accumulation of air over 25-35 0 latitudes giving rise to high pressure belt region of descending air is known as Hadley cell.

Westerlies wind: There situated at about 600 – 650 latitudes a low pressure area in both the hemisphere is known as sub-polar low or polar front. The winds that flow from sub-tropical high pressure area (Located at 250 - 350 latitude in both the hemisphere) to the low pressure area, situated at about 600 - 650 latitude in both the hemisphere, are known as Westerlies or prevailing wisterias (anti trade winds) In the upper atmosphere the reverse air movement takes place. This circulation is known as feral cell. These winds instead flowing in straight line are deflected due to corollas force. In northern hemisphere their direction is North – West and in southern hemisphere it is South-West.

Polar winds or Polar Easterlies winds: Near the poles due to shrinkage of air and due to cooling, there exists permanent high pressure on the poles. Therefore winds flow from the polar high to sub-polar low pressure area at about 60-650 latitude. The wind flow in North-East direction in northern hemisphere and in south-East direction in southern hemisphere. These winds consist of clod air. The air circulation is known as polar cell.

Local Winds: These winds are generated due to local condition and hence influence over very small area, therefore such winds are called local wind.

Land and sea breeze: An interchange of air between the sea and coastal land due to unequal heating and cooling is known as land and sea breezes. They are local in nature. During day time the coastal land and sea breezes.. They are local in nature. During day time the coastal land as heated very fast as compared to sea water causing low pressure over the land.

Therefore the surface air blows from sea to land and this is known as sea breeze. While during night time, the land cools faster than the sea, causing high pressure area over land as compared to sea. Therefore air blows from land to sea and this is known as land breeze.

Mountain and valley breeze: An interchange of air between the mountain and valley due to unequal heating and cooling of the two places is known as mountain and valley breeze. During daytime, the valley breeze. During daytime, the valley floors become more heated; the air over it expands and rises.

This rising air slides up the mountain slope and is known as valley breeze. During night reverse process takes place. Due to cooling of the air in the valley contracts and consequently augmented by air from the neighbouring hills and mountains. The air on the mountain slopes also cools and slides down into the valley which is known as mountain breeze.

Ketabatic winds: A mass of cold air over an elevated plateau during the winter tends to become more dense through radioactive cooling and then will drain down the slopes into the valleys below. The resulting down slope, drainage type winds are called Ketabatic winds. Most are relatively gentle breezes, not exceeding 4 to 5 m/s occasionally how ever, the cold dense air may be set in motion by a migrating cyclone or anticyclone and the Ketabatic winds may then attain destructive violence.

Foehn Wind

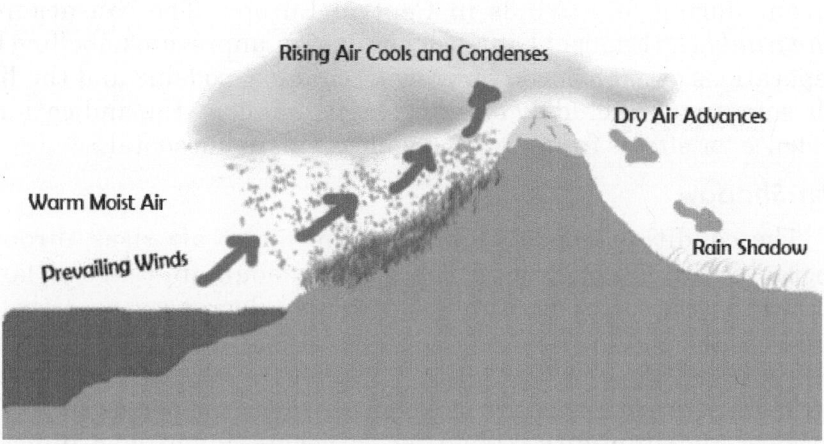

Figure: *How Föhn is produced*

A Föhn or Foehn is a type of dry, warm, down-slope wind that occurs in the lee (downwind side) of a mountain range.

It is a rain shadow wind that results from the subsequent adiabatic warming of air that has dropped most of its moisture on windward slopes. As a consequence of the different adiabatic lapse rates of moist and dry air, the air on the leeward slopes becomes warmer than equivalent elevations on the windward slopes. Föhn winds can raise temperatures by as much as 32 °C (58 °F) in just a matter of hours. Central Europe enjoys a warmer climate due to the Föhn, as moist winds off the Mediterranean Sea blow over the Alps.

Effects

Winds of this type are also called "snow-eaters" for their ability to make snow melt or sublimate rapidly. This snow-removing ability is caused not only by warmer temperatures, but also the low relative humidity of the air mass having been stripped of moisture by orographic precipitation coming over the mountain(s).

Föhn winds are notorious among mountaineers in the Alps, especially those climbing the Eiger, for whom the winds add further difficulty in ascending an already difficult peak.

They are also associated with the rapid spread of wildfires, making some regions which experience these winds particularly fire-prone.

Anecdotally, residents in areas of frequent föhn winds report illnesses ranging from migraines to psychosis. The first clinical review of these effects was published by the Austrian physician, Anton Czermak in the 19th century. A study by the Ludwig-Maximilians-Universität München found that suicide and accidents increased by 10 percent during föhn winds in Central Europe. The causation of *Föhnkrankheit* (English: Föhn-sickness) is yet unproven. Labelling for preparations of aspirin combined with caffeine, codeine and the like will sometimes include *Föhnkrankheit* amongst the indications. Evidence for effects from Chinook winds remain anecdotal.

Rain Shadow

The condition exists because warm moist air rises through orographic lifting up and over the top of a mountain range or large mountain. Because atmospheric pressure decreases as altitude increases, the air expands and cools adiabatically at the dry adiabatic lapse rate to the point that the air reaches its adiabatic dew point. Dew point is the temperature at which condensation begins at a given pressure and humidity. (Weather forecasts sometimes indicate humidity levels by giving a dew point at ground-level pressure for a given location, which says at what temperature dew will start condensing on the ground.)

Upon reaching the adiabatic dew point, water vapour in the air begins to condense, with the release of latent heat from condensation slowing the overall rate of adiabatic cooling of the air to the saturated adiabatic lapse rate as the air continues to rise. Condensation is also commonly followed by precipitation on the top and windward sides of the mountain. As the air descends on the leeward side, it is warmed by adiabatic compression at the dry adiabatic lapse rate. Because the air has lost much of its original water vapour content, the descending air creates an arid region on the leeward side of the mountain.

Etymology

The name *Foehn* arose in the Alpine region. Originating from Latin *(ventus) favonius*, a mild west wind of which Favonius was the Roman personification and probably transmitted by Romansh: *favuogn* or just *fuogn*, the term was adopted as Old High German: *phônno*. In the Southern Alps, the phenomenon is known as Italian: *favonio* and Slovene: *fen*. The German word "Fön" (without the "H", but pronounced the same way), a genericized trademark, is also used to mean "hairdryer," and the form "foehn" is used in Suisse Romande to mean "hairdryer" as well.

Local Examples

Regionally, these winds are known by many different names. These include:

- Föhn in Austria, southern Germany, Switzerland, France, Liechtenstein, and German-speaking regions of Northern Italy
- Feun in Piedmont
- Favonio in Ticino and Italy
- Fen in NW Slovenia
- Bergwind in South Africa
- Loo in Indo-Gangetic Plain
- Chinook winds east of the Rocky Mountains and the Cascade Range in the United States and Canada, and north, east and west of the Chugach Mountains of Alaska, United States
- Fogony in the Catalan Pyrenees
- Föhn in Wollongong and South Coast, NSW, Australia. Often associated with heavy orographic lifiting on the windward side of the escarpment
- Föhn in Ostrobothnia and Western Lapland in Finland as moist air crosses Scandinavian Mountains and dries up.

- Garmesh, Garmij, Garmbaad (Warm Wind): in Gilan region, in the south west of Caspian Sea in Iran
- Halny in the Carpathian Mountains, Poland (Central Europe)
- North-East Scotland, south-westerly winds create a Föhn effect bringing relatively warm temperatures on the lee side of the Grampians and Cairngorm mountain ranges. The reverse occurs when south-easterly winds create a Föhn effect to the North-West of Scotland, with the air drying out and warming up as it crosses the Grampians and Cairngorms from east to west. With the prevailing wind direction in the UK being from the west or south west, the Föhn effect in Scotland is more common in the North-East of the country, with the west of Scotland being much wetter.
- The Helm Wind, on the Pennines in the Eden Valley, Cumbria, England
- Hnúkaþeyr in Icelandic
- Lyvas wind in Thessalian plain, Voiotia plain, Plain of Thessaloniki, Elefsina and Athens in Greece
- The Nor'wester in Hawkes Bay, Canterbury, and Otago, New Zealand
- Puelche wind in Chile
- Terral in Málaga (southern Spain)
- Vântul Mare in the Carpathian Mountains, Romania
- Viento del Sur (Southern Wind) in the Cantabrian region (northern Spain)
- Zonda winds in Argentina
- Nortada in Cascais, and most notoriously in Guincho Beach, making it one of the best windsurfing spots in Europe
- Wuhan in China is famously known as one of the Three Furnaces on account of its extremely hot weather in summer resulting from the adiabatic warming effect created by mountains further south.

The Santa Ana winds of southern California, including the Sundowner winds of Santa Barbara, are in some ways similar to the Föhn, but originate in dry deserts as a katabatic wind.

Monsoon

Monsoon is traditionally defined as a seasonal reversing wind accompanied by corresponding changes in precipitation, but is now used to describe seasonal changes in atmospheric circulation and precipitation associated with the asymmetric heating of land and sea. Usually, the term monsoon is used to refer to the rainy phase of a seasonally-changing pattern, although technically there is also a dry phase.

The major monsoon systems of the world consist of the West African and Asia-Australian monsoons. The inclusion of the North and South American monsoons with incomplete wind reversal has been debated.

The term was first used in English in British India (now India, Bangladesh and Pakistan) and neighbouring countries to refer to the big seasonal winds blowing from the Bay of Bengal and Arabian Sea in the southwest bringing heavy rainfall to the area. The south-west monsoon winds are called 'Nairutya Maarut' in India.

Etymology

The English *monsoon* came from Portuguese *monηão*, ultimately from Arabic *mawsim*, "perhaps partly via early modern Dutch *monsun*".

History

Strengthening of the Asian monsoon has been linked to the uplift of the Tibetan Plateau after the collision of the Indian sub-continent and Asia around 50 million years ago. Because of studies of records from the Arabian Sea and that of the wind-blown dust in the Loess Plateau of China, many geologists believe the monsoon first became strong around 8 million years ago.

More recently, studies of plant fossils in China and new long-duration sediment records from the South China Sea led to a timing of the monsoon beginning 15–20 million years ago and linked to early Tibetan uplift. Testing of this hypothesis awaits deep ocean sampling by the Integrated Ocean Drilling Programme. The monsoon has varied significantly in strength since this time, largely linked to global climate change, especially the cycle of the Pleistocene ice ages. A study of marine plankton suggested that the Indian Monsoon strengthened around 5 million years ago.

Then, during ice periods, the sea level fell and the Indonesian Seaway closed. When this happened, cold waters in the Pacific were impeded from flowing into the Indian Ocean. It is believed that the resulting increase in sea surface temperatures in the Indian Ocean increased the intensity of monsoons.

Five episodes during the Quaternary at 2.22 Ma (PL-1), 1.83 Ma (PL-2), 0.68 Ma (PL-3), 0.45 Ma (PL-4) and 0.04 Ma (PL-5) were identified which showed a weakening of Leeuwin Current (LC). The weakening of the LC would have an effect on the sea surface temperature (SST) field in the Indian Ocean, as the Indonesian through flow generally warms the Indian Ocean. Thus these five intervals could probably be those of considerable lowering of SST in the Indian Ocean and would have influenced Indian monsoon intensity.

During the weak LC, there is the possibility of reduced intensity of the Indian winter monsoon and strong summer monsoon, because of change in the Indian Ocean dipole due to reduction in net heat input to the Indian Ocean through the Indonesian through flow. Thus a better understanding of the possible links between El Niño, Western Pacific Warm Pool, Indonesian Throughflow, wind pattern off western Australia, and ice volume expansion and contraction can be obtained by studying the behaviour of the LC during Quaternary at close stratigraphic intervals.

Strength of Impact

The impact of monsoon on the local weather is different from place to place. In some places there is just a likelihood of having a little more or less rain. In other places, quasi semi-deserts are turned into vivid green grasslands where all sorts of plants and crops can prosper.

The Indian Monsoon turns large parts of India from a kind of semi-desert into green lands. In places like this it is crucial for farmers to have the right timing for putting the seeds on the fields, as it is essential to use all the rain that is available for growing crops.

Process

Monsoons are large-scale sea breezes which occur when the temperature on land is significantly warmer or cooler than the temperature of the ocean. These temperature imbalances happen because oceans and land absorb heat in different ways. Over oceans, the air temperature remains relatively stable for two reasons: water has a relatively high heat capacity (3.9 to 4.2 J g^{-1} K^{-1}), and because both conduction and convection will equilibrate a hot or cold surface with deeper water (up to 50 metres). In contrast, dirt, sand, and rocks have lower heat capacities (0.19 to 0.35 J g^{-1} K^{-1}), and they can only transmit heat into the earth by conduction and not by convection. Therefore, bodies of water stay at a more even temperature, while land temperature are more variable.

During warmer months sunlight heats the surfaces of both land and oceans, but land temperatures rise more quickly. As the land's surface becomes warmer, the air above it expands and an area of low pressure develops. Meanwhile, the ocean remains at a lower temperature than the land, and the air above it retains a higher pressure. This difference in pressure causes sea breezes to blow from the ocean to the land, bringing moist air inland.

This moist air rises to a higher altitude over land and then it flows back toward the ocean (thus completing the cycle). However, when the air rises, and while it is still over the land, the air cools. This decreases the air's ability to hold water, and this causes precipitation over the land. This is why summer monsoons cause so much rain over land.

In the colder months, the cycle is reversed. Then the land cools faster than the oceans and the air over the land has higher pressure than air over the ocean. This causes the air over the land to flow to the ocean. When humid air rises over the ocean, it cools, and this causes precipitation over the oceans. (The cool air then flows towards the land to complete the cycle.)

Most summer monsoons have a dominant westerly component and a strong tendency to ascend and produce copious amounts of rain (because of the condensation of water vapour in the rising air). The intensity and duration, however, are not uniform from year to year. Winter monsoons, by contrast, have a dominant easterly component and a strong tendency to diverge, subside and cause drought.

Similar rainfall is caused when moist ocean air is lifted upwards by mountains, surface heating, convergence at the surface, divergence aloft, or from storm-produced outflows at the surface. However the lifting occurs, the air cools due to expansion in lower pressure, and this produces condensation.

Global Monsoons

Africa

The monsoon of western Sub-Saharan Africa is the result of the seasonal shifts of the Intertropical Convergence Zone and the great seasonal temperature and humidity differences between the Sahara and the equatorial Atlantic Ocean. It migrates northward from the equatorial Atlantic in February, reaches western Africa on or near June 22, then moves back to the south by October. The dry, northeasterly trade winds, and their more extreme form, the harmattan, are interrupted by the northern shift in the ITCZ and resultant southerly, rain-bearing winds during the summer. The semiarid Sahel and Sudan depend upon this pattern for most of their precipitation.

North America

The North American monsoon (NAM) occurs from late June or early July into September, originating over Mexico and spreading into the southwest United States by mid-July. It affects Mexico along the Sierra Madre Occidental as well as Arizona, New Mexico, Nevada, Utah, Colourado, West Texas and California. It pushes as far west as the Peninsular Ranges and Transverse Ranges of Southern California, but rarely reaches the coastal strip (a wall of desert thunderstorms only a half-hour's drive away is a common summer sight from the sunny skies along the coast during the monsoon). The North American monsoon is known to many as the *Summer*, *Southwest*, *Mexican* or *Arizona* monsoon. It is also sometimes called the *Desert monsoon* as a large part of the affected area are the Mojave and Sonoran deserts. However, it is debatable whether the North and South American weather patterns with incomplete wind reversal should be counted as true monsoons.

Asia

The Asian monsoons may be classified into a few sub-systems, such as the South Asian Monsoon which affects the Indian subcontinent and surrounding regions, and the East Asian Monsoon which affects southern China, Korea and parts of Japan.

South Asian Monsoon

Southwest monsoon: The southwestern summer monsoons occur from July through September. The Thar Desert and adjoining areas of the northern and central Indian subcontinent heats up considerably during the hot summers. This causes a low pressure area over the northern and central Indian subcontinent. To fill this void, the moisture-laden winds from the Indian Ocean rush in to the subcontinent. These winds, rich in moisture, are drawn towards the Himalayas. The Himalayas act like a high wall, blocking the winds from passing into Central Asia, and forcing them to rise. As the clouds rise their temperature drops and precipitation occurs. Some areas of the subcontinent receive up to 10,000 mm (390 in) of rain annually.

The southwest monsoon is generally expected to begin around the beginning of June and fade away by the end of September. The moisture-laden winds on reaching the southernmost point of the Indian Peninsula, due to its topography, become divided into two parts: the *Arabian Sea Branch* and the *Bay of Bengal Branch*.

The *Arabian Sea Branch* of the Southwest Monsoon first hits the Western Ghats of the coastal state of Kerala, India, thus making this area the first state in India to receive rain from the Southwest Monsoon. This branch of the monsoon moves northwards along the Western Ghats (Konkan and Goa) with precipitation on coastal areas, west of the Western Ghats. The eastern areas of the Western Ghats do not receive much rain from this monsoon as the wind does not cross the Western Ghats.

The *Bay of Bengal Branch* of Southwest Monsoon flows over the Bay of Bengal heading towards North-East India and Bengal, picking up more moisture from the Bay of Bengal. The winds arrive at the Eastern Himalayas with large amounts of rain. Mawsynram, situated on the southern slopes of the Khasi Hills in Meghalaya, India, is one of the wettest places on Earth. After the arrival at the Eastern Himalayas, the winds turns towards the west, travelling over the Indo-Gangetic Plain at a rate of roughly 1–2 weeks per state, pouring rain all along its way. June 1 is regarded as the date of onset of the monsoon in India, as indicated by the arrival of the monsoon in the southernmost state of Kerala.

The monsoon accounts for 80% of the rainfall in India. Indian agriculture (which accounts for 25% of the GDP and employs 70% of the population) is heavily dependent on the rains, for growing crops especially like cotton, rice, oilseeds and coarse grains. A delay of a few days in the arrival of the monsoon can badly affect the economy, as evidenced in the numerous droughts in India in the 1990s.

The monsoon is widely welcomed and appreciated by city-dwellers as well, for it provides relief from the climax of summer heat in June. However, the roads take a battering every year. Often houses and streets are waterlogged and slums are flooded despite drainage systems. A lack of city infrastructure coupled with changing climate patterns causes severe economical loss including damage to property and loss of lives, as evidenced in the 2005 flooding in Mumbai that brought Mumbai to a standstill. Bangladesh and certain regions of India like Assam and West Bengal, also frequently experience heavy floods during this season. Recently, areas in India that used to receive scanty rainfall throughout the year, like the Thar Desert, have surprisingly ended up receiving floods due to the prolonged monsoon season.

The influence of the Southwest Monsoon is felt as far north as in China's Xinjiang. It is estimated that about 70% of all precipitation in the central part of the Tian Shan Mountains falls during the three summer months, when the region is under the monsoon influence; about 70% of that is directly of "cyclonic" (i.e., monsoon-driven) origin (as opposed to "local convection").

Northeast Monsoon

Around September, with the sun fast retreating south, the northern land mass of the Indian subcontinent begins to cool off rapidly. With this air pressure begins to build over northern India, the Indian Ocean and its surrounding atmosphere still holds its heat. This causes cold wind to sweep down from the Himalayas and Indo-Gangetic Plain towards the vast spans of the Indian Ocean south of the Deccan peninsula. This is known as the Northeast Monsoon or Retreating Monsoon.

While travelling towards the Indian Ocean, the dry cold wind picks up some moisture from the Bay of Bengal and pours it over peninsular India and parts of Sri Lanka. Cities like Chennai, which get less rain from the Southwest Monsoon, receives rain from this Monsoon. About 50% to 60% of the rain received by the state of Tamil Nadu is from the Northeast Monsoon. In Southern Asia, the northeastern monsoons take place from December to early March when the surface high-pressure system is strongest. The jet stream in this region splits into the southern subtropical jet and the polar jet. The subtropical flow directs northeasterly winds to blow across southern Asia, creating dry air streams which produce clear skies over India. Meanwhile, a low pressure system develops over South-East Asia and Australasia and winds are directed toward Australia known as a monsoon trough.

East Asian Monsoon

The East Asian monsoon affects large parts of Indo-China, Philippines, China, Korea and Japan. It is characterised by a warm, rainy summer monsoon and a cold, dry winter monsoon. The rain occurs in a concentrated belt that stretches east-west except in East China where it is tilted east-northeast over Korea and Japan. The seasonal rain is known as *Meiyu* in China, *Changma* in Korea, and *Bai-u* in Japan, with the latter two resembling frontal rain.

The onset of the summer monsoon is marked by a period of premonsoonal rain over South China and Taiwan in early May. From May through August, the summer monsoon shifts through a series of dry and rainy phases as the rain belt moves northward, beginning over Indochina and the South China Sea (May), to the Yangtze River Basin and Japan (June) and finally to North China and Korea (July). When the monsoon ends in August, the rain belt moves back to South China.

Australia

Also known as the Indo-Australian Monsoon. The rainy season occurs from September to February and it is a major source of energy for the Hadley circulation during boreal winter. The *Maritime Continent Monsoon* and the *Australian Monsoon* may be considered to be the same system, the Indo-Australian Monsoon.

It is associated with the development of the Siberian High and the movement of the heating maxima from the Northern Hemisphere to the Southern Hemisphere. North-easterly winds flow down Southeast Asia, are turned north-westerly/westerly by Borneo topography towards Australia. This forms a cyclonic circulation vortex over Borneo, which together with descending cold surges of winter air from higher latitudes, cause significant weather phenomena in the region. Examples are the formation of a rare low-latitude tropical storm in 2001, Tropical Storm Vamei, and the devastating flood of Jakarta in 2007.

The onset of the monsoon over the Maritime Continent tends to follow the heating maxima down Vietnam and the Malay Peninsula (September), to Sumatra, Borneo and the Philippines (October), to Java, Sulawesi (November), Irian Jaya and Northern Australia (December, January). However, the monsoon is not a simple response to heating but a more complex interaction topography, wind and sea, as demonstrated by its abrupt rather than gradual withdrawal from the region. The Australian monsoon (the "Wet") occurs in the southern summer when the monsoon trough develops over Northern Australia. Over three-quarters of annual rainfall in Northern Australia falls during this time.

Europe

The European Monsoon (more commonly known as the Return of the Westerlies) is the result of a resurgence of westerly winds from the Atlantic, where they become loaded with wind and rain. These Westerly winds are a common phenomenon during the European winter, but they ease as spring approaches in late March and through April and May. The winds pick up again in June, which is why this phenomenon is also referred to as "the return of the westerlies".

The rain usually arrives in two waves, at the beginning of June and again in mid to late June. The European monsoon is not a monsoon in the traditional sense in that it doesn't meet all the requirements to be classified as such.

Instead the Return of the Westerlies is more regarded as a conveyor belt that delivers a series of low pressure centres to Western Europe where they create unsettled weather. These storms generally feature significantly lower than average temperatures, fierce rain or hail, thunder and strong winds.

The Return of the Westerlies affects Europe's Northern Atlantic coastline, more precisely Ireland, Great Britain, the Benelux countries, Western Germany, Northern France and parts of Scandinavia.

Monsoon Wind

An interchange of air between the land and oceans due to unequal heating and cooling of continents and oceans is known as monsoon winds.

It has an annual period of occurrence. During summer, the land is heated Very much as compared to the oceans which cause which oceans low pressure Over the land and the winds blow from the oceans to the continents. During winter, land cools down faster than the oceans causing high Pressure over continents and low pressure over the oceans and the wind Blows from continents to oceans. The Indian monsoon is the best known Example of this alternating circulation system. There are two types of Monsoons over India i.e. south – west monsoon and North- East monsoon.

South – West Monsoon: India is positional situated in North – East trade winds and should have N- E winds throughout the year, but a low pressure through lies along the Ganges and upper India, due to which S.- W winds predominate. During April to September a low pressure centre is formed over N – W India. The s-W trade winds of the Indian ocean blow to the equator and then turning to the right under carioles force and move on a S – W winds Around the low pressure centre over India. [This monsoon blows from the African coast (150E)]. The moisture laden air while rising the mountain of Asia cools, condense and precipitate. As a result the pressure is lowered to increase the pressure gradient.

North – East Monsoon: Complete reversal of the S – W monsoon winds takes place as the high pressure centre is located in eastern Asian (1035 mb) and low in about 1010 mb. During this time from North to South the cold season is established. This monsoon is active during October and November. The winds flow in North- East direction. This wind is generally dry but gives rains to AP, TN states. Monsoon winds also exist over west Africa, Brazil, eastern USA, Australia. Philippines etc.

Forces Acting to Produce Wind

Wind is motion of air in response to unbalanced forces acting in horizontal direction. The different forces involved in the flow of the wind are described below:-

Horizontal pressure gradient (PG) force: The rate of change in atmospheric pressure between two points at the Same elevation is called the pressure gradient or isobaric slope. It is proportional to the difference in pressure and is the immediate Cause of horizontal air movement. The direction of air flow is from high to Low pressure and the speed of flow is directly related to the pressure Gradient. The pressure gradient is said tube steep when the rate of change Is great and the gradient the more rapid will be the flow of air. The direction of the pressure gradient is perpendicular to the isobars and pointing towards low pressure.

PG = 1/P* dp/dn

Where P= air density

Dp/dn = rate of change in pressure with distance.

The earth's rotational deflective force [Coriolis force]: This force comes into play due to rotation of the earth on it axis. It has most potent influences upon wind direction. The Coriolis force effect Causes all winds in the nor them hemisphere to move or deflect toward the Right and those of the sot herm hemisphere to move to the left with respect to the rotating earth. At the equator the effect has a value of zero (Sin 90=1).and it increases regularly towards the poles (and becomes Sin 90=1). The Coriolis effect changes wind direction but does not change wind speed.

Centrifugal force: This force tends to throw the air particles outward from the centre of small circle path on which the particle is moving. The centrifugal force works against the gravitational attraction directed forwards the earth centre.

Frictional force: The roughness of the surface provides frictional resistance to the air motion. It is the retarding effect of trees, buildings and other irregularities in the topography; It is always opposed to the direction of the air motion and therefore fends to decrease the wind speed. Friction causes a movement of air across the isobars towards low pressure.

Geotropic winds: When a wind flows in a straight line with no acceleration or frictional force on it, the only forces acting are the carioles and pressure gradient foresee. The wind that blows under these conditions is called as geotropic wind.

Cyclone

In meteorology, a cyclone is an area of closed, circular fluid motion rotating in the same direction as the Earth. This is usually characterized by inward spiraling winds that rotate counter-clockwise in the Northern

Hemisphere and clockwise in the Southern Hemisphere of the Earth. Most large-scale cyclonic circulations are centred on areas of low atmospheric pressure. The largest low-pressure systems are cold-core polar cyclones and extratropical cyclones which lie on the synoptic scale. According to the National Hurricane Centre glossary, warm-core cyclones such as tropical cyclones and subtropical cyclones also lie within the synoptic scale. Mesocyclones, tornadoes and dust devils lie within the smaller mesoscale. Upper level cyclones can exist without the presence of a surface low, and can pinch off from the base of the Tropical Upper Tropospheric Trough during the summer months in the Northern Hemisphere. Cyclones have also been seen on extraterrestrial planets, such as Mars and Neptune. Cyclogenesis describes the process of cyclone formation and intensification. Extratropical cyclones form as waves in large regions of enhanced mid-latitude temperature contrasts called baroclinic zones. These zones contract to form weather fronts as the cyclonic circulation closes and intensifies. Later in their life cycle, cyclones occlude as cold core systems. A cyclone's track is guided over the course of its 2 to 6 day life cycle by the steering flow of the cancer or subtropical jet stream.

Weather fronts separate two masses of air of different densities and are associated with the most prominent meteorological phenomena. Air masses separated by a front may differ in temperature or humidity. Strong cold fronts typically feature narrow bands of thunderstorms and severe weather, and may on occasion be preceded by squall lines or dry lines. They form west of the circulation centre and generally move from west to east. Warm fronts form east of the cyclone centre and are usually preceded by stratiform precipitation and fog. They move poleward ahead of the cyclone path. Occluded fronts form late in the cyclone life cycle near the centre of the cyclone and often wrap around the storm centre.

Tropical cyclogenesis describes the process of development of tropical cyclones. Tropical cyclones form due to latent heat driven by significant thunderstorm activity, and are warm core. Cyclones can transition between extratropical, subtropical, and tropical phases under the right conditions. Mesocyclones form as warm core cyclones over land, and can lead to tornado formation. Waterspouts can also form from mesocyclones, but more often develop from environments of high instability and low vertical wind shear. In the Atlantic and the northeastern Pacific oceans, a tropical cyclone is generally referred to as a hurricane (from the name of the ancient Central American deity of wind, Huracan), in the Indian and south Pacific oceans it is called a cyclone, and in the northwestern Pacific it is called a typhoon.

Structure

There are a number of structural characteristics common to all cyclones. A cyclone is a low-pressure area. A cyclone's centre (often known in a mature tropical cyclone as the eye), is the area of lowest atmospheric pressure in the region. Near the centre, the pressure gradient force (from the pressure in the centre of the cyclone compared to the pressure outside the cyclone) and the force from the Coriolis effect must be in an approximate balance, or the cyclone would collapse on itself as a result of the difference in pressure.

Because of the Coriolis effect, the wind flow around a large cyclone is counterclockwise in the Northern Hemisphere and clockwise in the Southern Hemisphere. Cyclonic circulation is sometimes referred to as contra solem. In the Northern Hemisphere, the fastest winds relative to the surface of the Earth therefore occur on the eastern side of a northward-moving cyclone and on the northern side of a westward-moving one; the opposite occurs in the Southern Hemisphere. (The wind flow around an anticyclone, on the other hand, is clockwise in the northern hemisphere, and counterclockwise in the southern hemisphere.)

Formation

Cyclogenesis is the development or strengthening of cyclonic circulation in the atmosphere (a low-pressure area). Cyclogenesis is an umbrella term for several different processes, all of which result in the development of some sort of cyclone. It can occur at various scales, from the microscale to the synoptic scale.

Extratropical cyclones form as waves along weather fronts before occluding later in their life cycle as cold core cyclones.

Tropical cyclones form due to latent heat driven by significant thunderstorm activity, and are warm core.

Mesocyclones form as warm core cyclones over land, and can lead to tornado formation. Waterspouts can also form from mesocyclones, but more often develop from environments of high instability and low vertical wind shear. Cyclogenesis is the opposite of cyclolysis, and has an anticyclonic (high-pressure system) equivalent which deals with the formation of high-pressure areas—Anticyclogenesis.

The surface low has a variety of ways of forming. Topography can force a surface low when dense low-level high-pressure system ridges in east of a north-south mountain barrier. Mesoscale convective systems can spawn surface lows which are initially warm core. The disturbance can grow into a wave-like formation along the front and the low will be positioned at the crest. Around the low, flow will become cyclonic, by definition. This rotational flow will push polar air equatorward west of the low via its trailing cold front, and warmer air with push poleward low via the warm front. Usually the cold front will move at a quicker pace than the warm front and "catch up" with it due to the slow erosion of higher density airmass located out ahead of the cyclone and the higher density airmass sweeping in behind the cyclone, usually resulting in a narrowing warm sector. At this point an occluded front forms where the warm air mass is pushed upwards into a trough of warm air aloft, which is also known as a trowal.

Tropical cyclogenesis is the technical term describing the development and strengthening of a tropical cyclone in the atmosphere. The mechanisms through which tropical cyclogenesis occurs are distinctly different from those through which mid-latitude cyclogenesis occurs. Tropical cyclogenesis involves the development of a warm-core cyclone, due to significant convection in a favourable atmospheric environment. There are six main requirements for tropical cyclogenesis: sufficiently warm sea surface temperatures, atmospheric instability, high humidity in the lower to middle levels of the troposphere, enough Coriolis force to develop a low-pressure centre, a preexisting low-level focus or disturbance, and low vertical wind shear. An average of 86 tropical cyclones of tropical storm intensity form annually worldwide, with 47 reaching hurricane/ typhoon strength, and 20 becoming intense tropical cyclones (at least Category 3 intensity on the Saffir–Simpson Hurricane Scale).

Synoptic Scale

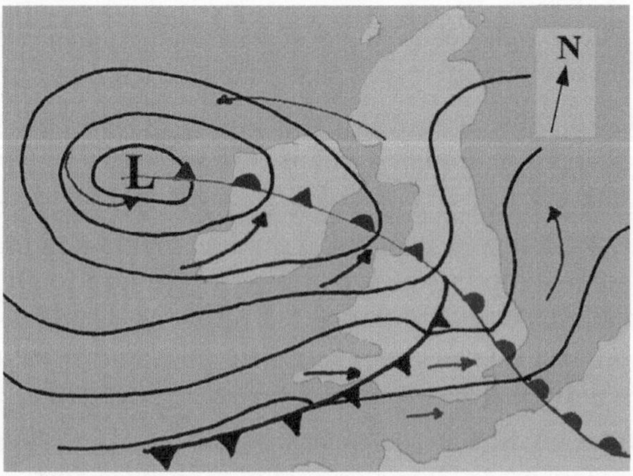

Figure: *A fictitious synoptic chart of an extratropical cyclone affecting the UK and Ireland. The blue arrows between isobars indicate the direction of the wind, while the "L" symbol denotes the centre of the "low". Note the occluded, cold and warm frontal boundaries.*

The following types of cyclones are identifiable in synoptic charts.

Surface-based Types

There are three main types surface-based cyclones: Extratropical cyclones, Subtropical cyclones and Tropical cyclones

Extratropical Cyclone

An extratropical cyclone is a synoptic scale low-pressure weather system that does not have tropical characteristics, being connected with fronts and horizontal gradients in temperature and dew point otherwise known as "baroclinic zones".

The descriptor "extratropical" refers to the fact that this type of cyclone generally occurs outside of the tropics, in the middle latitudes of the planet. These systems may also be described as "mid-latitude cyclones" due to their area of formation, or "post-tropical cyclones" where extratropical transition has occurred, and are often described as "depressions" or "lows" by weather forecasters and the general public. These are the everyday phenomena which along with anti-cyclones, drive the weather over much of the Earth.

Although extratropical cyclones are almost always classified as baroclinic since they form along zones of temperature and dewpoint gradient within the westerlies, they can sometimes become barotropic late in their life cycle when the temperature distribution around the cyclone becomes fairly uniform with radius. An extratropical cyclone can transform into a subtropical storm, and from there into a tropical cyclone, if it dwells over warm waters and develops central convection, which warms its core. One intense type of extratropical cyclone that strikes during wintertime is a *nor'easter*.

Polar Low

A polar low is a small-scale, short-lived atmospheric low-pressure system (depression) that is found over the ocean areas poleward of the main polar front in both the Northern and Southern Hemispheres. Polar lows are cold-core so they can be considered as a subset of extratropical cyclones. Polar lows were first identified on the meteorological satellite imagery that became available in the 1960s, which revealed many small-scale cloud vortices at high latitudes. The most active polar lows are found over certain ice-free maritime areas in or near the Arctic during the winter, such as the Norwegian Sea, Barents Sea, Labrador Sea and Gulf of Alaska. Polar lows dissipate rapidly when they make landfall. Antarctic systems tend to be weaker than their northern counterparts since the air-sea temperature differences around the continent are generally smaller. However, vigorous polar lows can be found over the Southern Ocean. During winter, when cold-core lows with temperatures in the mid-levels of the troposphere reach –45 °C (–49 °F) move over open waters, deep convection forms which allows polar low development to become possible.

The systems usually have a horizontal length scale of less than 1,000 kilometres (620 mi) and exist for no more than a couple of days. They are part of the larger class of mesoscale weather systems. Polar lows can be difficult to detect using conventional weather reports and are a hazard to high-latitude operations, such as shipping and gas and oil platforms. Polar lows have been referred to by many other terms, such as polar mesoscale vortex, Arctic hurricane, Arctic low, and cold air depression. Today the term is usually reserved for the more vigorous systems that have near-surface winds of at least 17 m/s.

Subtropical

A subtropical cyclone is a weather system that has some characteristics of a tropical cyclone and some characteristics of an extratropical cyclone. They can form between the equator and the 50th parallel. As early as the 1950s, meteorologists were unclear whether they should be characterized as tropical cyclones or extratropical cyclones, and used terms such as quasi-tropical and semi-tropical to describe the cyclone hybrids. By 1972, the National Hurricane Centre officially recognized this cyclone category. Subtropical cyclones began to receive names off the official tropical cyclone list in the Atlantic Basin in 2002. They have broad wind patterns with maximum sustained winds located farther from the centre than typical tropical cyclones, and exist in areas of weak to moderate temperature gradient.

Since they form from initially extratropical cyclones which have colder temperatures aloft than normally found in the tropics, the sea surface temperatures required for their formation are lower than the tropical cyclone threshold by three degrees Celsius, or five degrees Fahrenheit, lying around 23 degrees Celsius. This means that subtropical cyclones are more likely to form outside the traditional bounds of the hurricane season. Although subtropical storms rarely have hurricane-force winds, they may become tropical in nature as their cores warm.

Tropical

A tropical cyclone is a storm system characterized by a low-pressure centre and numerous thunderstorms that produce strong winds and flooding rain. A tropical cyclone feeds on heat released when moist air rises, resulting in condensation of water vapour contained in the moist air. They are fuelled by a different heat mechanism than other cyclonic windstorms such as nor'easters, European windstorms, and polar lows, leading to their classification as "warm core" storm systems.The term "tropical" refers to both the geographic origin of these systems, which form almost exclusively in tropical regions of the globe, and their formation in Maritime Tropical air masses.

The term "cyclone" refers to such storms' cyclonic nature, with counterclockwise rotation in the Northern Hemisphere and clockwise rotation in the Southern Hemisphere. Depending on their location and strength, tropical cyclones are referred to by other names, such as hurricane, typhoon, tropical storm, cyclonic storm, tropical depression, or simply as a cyclone.

While tropical cyclones can produce extremely powerful winds and torrential rain, they are also able to produce high waves and damaging storm surge. They develop over large bodies of warm water, and lose their strength if they move over land. This is the reason coastal regions can receive significant damage from a tropical cyclone, while inland regions are relatively safe from receiving strong winds. Heavy rains, however, can produce significant flooding inland, and storm surges can produce extensive coastal flooding up to 40 kilometres (25 mi) from the coastline. Although their effects on human populations can be devastating, tropical cyclones can also relieve drought conditions. They also carry heat and energy away from the tropics and transport it toward temperate latitudes, which makes them an important part of the global atmospheric circulation mechanism. As a result, tropical cyclones help to maintain equilibrium in the Earth's troposphere.

Many tropical cyclones develop when the atmospheric conditions around a weak disturbance in the atmosphere are favourable. Others form when other types of cyclones acquire tropical characteristics. Tropical systems are then moved by steering winds in the troposphere; if the conditions remain favourable, the tropical disturbance intensifies, and can even develop an eye. On the other end of the spectrum, if the conditions around the system deteriorate or the tropical cyclone makes landfall, the system weakens and eventually dissipates. A tropical cyclone can become extratropical as it moves toward higher latitudes if its energy source changes from heat released by condensation to differences in temperature between air masses; From an operational standpoint, a tropical cyclone is usually not considered to become subtropical during its extratropical transition.

Upper Level Types

Polar Cyclone

A polar, sub-polar, or Arctic cyclone (also known as a polar vortex) is a vast area of low pressure which strengthens in the winter and weakens in the summer. A polar cyclone is a low-pressure weather system, usually spanning 1,000 kilometres (620 mi) to 2,000 kilometres (1,200 mi), in which the air circulates in a counterclockwise direction in the northern hemisphere, and a clockwise direction in the southern hemisphere. In the Northern Hemisphere, the polar cyclone has two centres on average. One centre lies near Baffin Island and the other over northeast Siberia. In the southern hemisphere, it tends to be located near the edge of the Ross ice shelf near 160 west longitude. When the polar vortex is strong, westerly flow descends to the Earth's surface. When the polar cyclone is weak, significant cold outbreaks occur.

TUTT Cell

Under specific circumstances, upper cold lows can break off from the base of the Tropical Upper Tropospheric Trough (TUTT), which is located mid-ocean in the Northern Hemisphere during the summer months. These upper tropospheric cyclonic vortices, also known as TUTT cells or TUTT lows, usually move slowly from east-northeast to west-southwest, and generally do not extend below 20,000 feet in altitude. A weak inverted surface trough within the trade wind is generally found underneath them, and they may also be associated with broad areas of high-level clouds. Downward development results in an increase of cumulus clouds and the appearance of a surface vortex. In rare cases, they become warm-core, resulting in the vortex becoming a tropical cyclone. Upper cyclones and upper troughs which trail tropical cyclones can cause additional outflow channels and aid in their intensification process. Developing tropical disturbances can help create or deepen upper troughs or upper lows in their wake due to the outflow jet emanating from the developing tropical disturbance/cyclone.

Mesoscale

The following types of cyclones are not identifiable in synoptic charts.

Mesocyclone

A mesocyclone is a vortex of air, 2.0 kilometres (1.2 mi) to 10 kilometres (6.2 mi) in diametre (the mesoscale of meteorology), within a convective storm. Air rises and rotates around a vertical axis, usually in the same direction as low-pressure systems in both northern and southern hemisphere. They are most often cyclonic, that is, associated with a localized low-pressure region within a supercell. Such storms can feature strong surface winds and severe hail. Mesocyclones often occur together with updrafts in supercells, where tornadoes may form. About 1700 mesocyclones form annually across the United States, but only half produce tornadoes.

Tornado

A tornado is a violently rotating column of air that is in contact with both the surface of the earth and a cumulonimbus cloud or, in rare cases, the base of a cumulus cloud. They are often referred to as twisters or cyclones, although the word cyclone is used in meteorology, in a wider sense, to name any closed low-pressure circulation.

Dust Devil

A dust devil is a strong, well-formed, and relatively long-lived whirlwind, ranging from small (half a metre wide and a few metres tall) to large (more than 10 metres wide and more than 1000 metres tall). The primary vertical motion is upward. Dust devils are usually harmless, but can on rare occasions grow large enough to pose a threat to both people and property.

Waterspout

A columnar vortex forming over water that is, in its most common form, a non-supercell tornado over water that is connected to a cumuliform cloud. While it is often weaker than most of its land counterparts, stronger versions spawned by mesocyclones do occur.

Steam Devil

A gentle vortex over calm water or wet land made visible by rising water vapour.

Other Planets

Cyclones are not unique to Earth. Cyclonic storms are common on Jovian planets, such as the Small Dark Spot on Neptune. It is about one third the diametre of the Great Dark Spot and received the nickname "Wizard's Eye" because it looks like an eye. This appearance is caused by a white cloud in the middle of the Wizard's Eye. Mars has also exhibited cyclonic storms. Jovian storms like the Great Red Spot are usually mistakenly named as giant hurricanes or cyclonic storms. However, this is inaccurate, as the Great Red Spot is, in fact, the inverse phenomenon, an anticyclone.

Anticyclone

An anticyclone (that is, opposite to a cyclone) is a weatherphenomenon defined by the United States National Weather Service's glossary as "[a] large-scale circulation of winds around a central region of high atmospheric pressure, clockwise in the Northern Hemisphere, counterclockwise in the Southern Hemisphere". Effects of surface-based anticyclones include clearing skies as well as cooler, drier air. Fog can also form overnight within a region of higher pressure.

Mid-tropospheric systems, such as the subtropical ridge, deflect tropical cyclones around their periphery and cause a temperature inversion inhibiting free convection near their centre, building up surface-based haze under their base. Anticyclones aloft can form within warm core lows such as tropical cyclones, due to descending cool air from the backside of upper troughs such as polar highs, or from large scale sinking such as the subtropical ridge.

History

Sir Francis Galton first discovered anticyclones in the 1860's. Preferred areas within a synoptic flow pattern in higher levels of the troposphere are beneath the western side of troughs, or dips in the Rossby wave pattern. High-pressure systems are alternatively referred to as anticyclones. Their circulation is sometimes referred to as cum sole. The subtropical ridge forms due to the Hadley cell circulation between the equator and the subtropics of the Northern Hemisphere and Southern Hemisphere.

Upper-level high-pressure areas lie over tropical cyclones due to their warm core nature.

Surface anticyclones form due to downward motion through the troposphere, the atmospheric layer where weather occurs. Preferred areas within a synoptic flow pattern in higher levels of the troposphere are beneath the western side of troughs. On weather maps, these areas show converging winds (isotachs), also known as confluence, or converging height lines near or above the level of non-divergence, which is near the 500 hPa pressure surface about midway up the troposphere. Because they weaken with height, these high-pressure systems are cold.

Subtropical Ridge

Heating of the earth near the equator forces upward motion and convection along the monsoon trough or intertropical convergence zone.

The divergence over the near-equatorial trough leads to air rising and moving away from the equator aloft. As air moves towards the mid-latitudes, it cools and sinks leading to subsidence near the 30° parallel of both hemispheres. This circulation known as the Hadley cell forms the subtropical ridge.

Many of the world's deserts are caused by these climatological high-pressure areas. Because these anticyclones strengthen with height, they are known as warm core ridges.

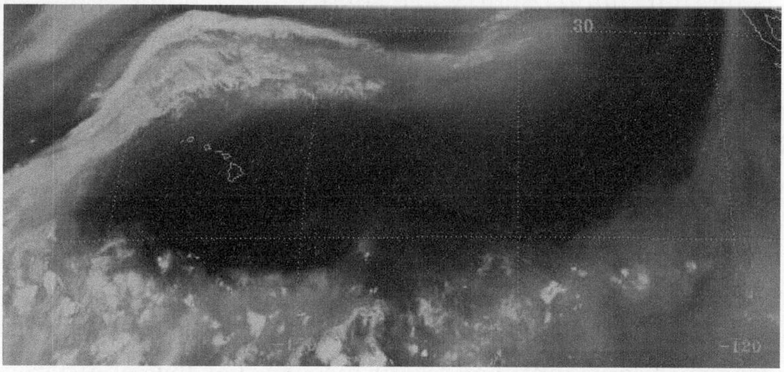

Figure: *The subtropical ridge shows up as a large area of black (dryness) on this water vapour satellite image from September 2000.*

Formation Aloft

The development of anticyclones aloft occurs in warm core cyclones such as tropical cyclones when latent heat caused by the formation of clouds is released aloft increasing the air temperature; the resultant thickness of the atmospheric layer increases high pressure aloft which evacuates their outflow.

Structure

In the absence of rotation, the wind tends to blow from areas of high pressure to areas of low pressure. The stronger the pressure difference (pressure gradient) between a high-pressure system and a low-pressure system, the stronger the wind. The coriolis force caused by Earth's rotation gives winds within high-pressure systems their clockwise circulation in the northern hemisphere (as the wind moves outward and is deflected right from the centre of high pressure) and anticlockwise circulation in the southern hemisphere (as the wind moves outward and is deflected left from the centre of high pressure). Friction with land slows down the wind flowing out of high-pressure systems and causes wind to flow more outward (more ageostrophically) from the centre.

Effects

Surface-based Systems

High-pressure systems are frequently associated with light winds at the surface and subsidence through the lower portion of the troposphere. Subsidence will generally dry out an air mass by adiabatic (compressional) heating. Thus, high pressure typically brings clear skies. Because no clouds are present to reflect sunlight during the day, there is more incoming shortwave solar radiation and temperatures rise. At night, the absence of clouds means that outgoing longwave radiation (i.e. heat energy from the surface) is not absorbed, giving cooler diurnal low temperatures in all seasons. When surface winds become light, the subsidence produced directly under a high-pressure system can lead to a buildup of particulates in urban areas under the ridge, leading to widespread haze. If the low level relative humidity rises towards 100 percent overnight, fog can form.

Strong but vertically shallow high-pressure systems moving from higher latitudes to lower latitudes in the northern hemisphere are associated with continental arctic air masses. The low, sharp inversion can lead to areas of persistent stratocumulus or stratus cloud, colloquially known as anticyclonic gloom. The type of weather brought about by an anticyclone depends on its origin. For example, extensions of the Azores high pressure may bring about anticyclonic gloom during the winter because they are warmed at the base and trap moisture as they move over the warmer oceans. High pressures that build to the north and extend southwards often bring clear weather because they are cooled at the base (as opposed to warmed) which helps prevent clouds from forming.

Once arctic air moves over an unfrozen ocean, the air mass modifies greatly over the warmer water and takes on the character of a maritime air mass, which reduces the strength of the high-pressure system. When extremely cold air moves over relatively warm oceans, polar lows can develop. However, warm and moist (or maritime tropical) air masses which move poleward from tropical sources are slower to modify than arctic air masses.

Mid-tropospheric Systems

The circulation around mid-level ridges, and the subsidence at their centre, act to steer tropical cyclones around their periphery. Due to the subsidence within this type of system, a *cap* can be set up which inhibits the development of free convection. This limits thunderstorm activity near their centre and traps low-level pollutants such as ozone as haze under their base, which is a significant problem in large urban centres during summer months such as Los Angeles, California and Mexico City.

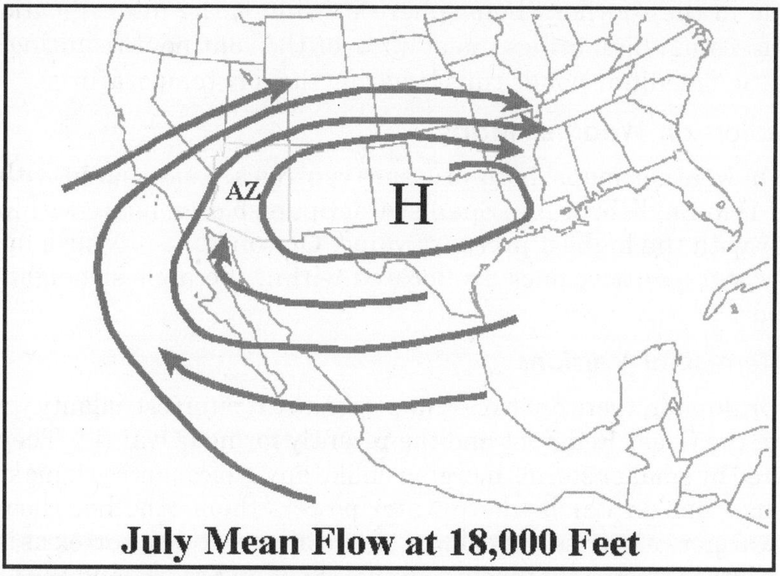

Figure: *Mean July subtropical ridge position in North America*

Upper Tropospheric Systems

The existence of an upper level ridge allows upper level divergence which leads to surface convergence. If a capping mid-level ridge does not exist, this leads to free convection and the development of showers and thunderstorms if the lower atmosphere is humid. Because tropical cyclones strengthen these ridges, a positive feedback loop develops between the convective tropical cyclone and the upper level high, strengthening both systems. This loop stops once ocean temperatures under the system cool sufficiently, under 26.5 °C (79.7 °F), which forces the thunderstorm activity to wane, which then weakens the upper level ridge.

Importance to Global Monsoon Regimes

When the subtropical ridge in the northwest Pacific is stronger than normal, it leads to a wet monsoon season for Asia. The subtropical ridge position is linked to how far northward monsoon moisture and thunderstorms extend into the United States. Typically, the subtropical ridge across North America migrates far enough northward to begin monsoon conditions across the Desert Southwest from July to

September. When the subtropical ridge is farther north than normal towards the Four Corners, monsoon thunderstorms can spread northward into Arizona. When suppressed to the south, the atmosphere dries out across the Desert Southwest, causing a break in the monsoon regime in the southern hemisphere in winter, the intensification of the mascaleni high brings moist air over the continent resulting into formation medium clouds thus lowering the day temperatures.

Depiction on Weather Maps

On weather maps, high-pressure centres are associated with the letter H in English, or A in Spanish (for Spanish *alta*: high), within the isobar with the highest pressure value. On constant-pressure upper-level charts, anticyclones are located within the highest height line contour.

Extraterrestrial Versions

On Jupiter, there are two examples of an extraterrestrialanticyclonic storm; the Great Red Spot and the recently formed Oval BA. They are powered by smaller storms merging unlike any typical anticyclonic storm that happens on Earth where water powers them. Another theory is that warmer gases rise in a column of cold air, creating a vortex as is the case of other storms that include Anne's Spot on Saturn, and the Great Dark Spot on Neptune. Anticyclones have been detected near the poles of Venus.

Variation in Relative Humidity

Diurnal variation in RH: The diurnal variation in relative humidity is approximately inverse to that of temperature about sunrise the RH is maximum and between 14 to 15 hrs minimum RH is observed.

Annual variation in RH: The annual variation of relative humidity is largely depends upon the locality.

At regions where the rainy season is in summer and winter is dry, the maximum relative humidity occurs in summer and minimum in winter and at other regions maximum RH occurs in winter. Over ocean RH reaches maximum in summer.

Variation of RH with Latitude and Altitude:

RH shows maximum at equation or about 80%. Therefore it decreases to 70% in the regions of high pressure belts in 30-350 and afterwards increases again to 80 to 90% in the polar region.

The vertical variation of RH is not governed by any exact law. In or near the clouds it is 100% but below or above it is different. The moist air masses carried out by advection at altitudes also change relative humidity.

Cloud

In meteorology, a cloud is a visible mass of liquiddroplets or frozen crystals made of water or various chemicals suspended in the atmosphere above the surface of a planetary body. These suspended particles are also known as aerosols and are studied in the cloud physics branch of meteorology.

Terrestrial cloud formation is the result of air in Earth's atmosphere becoming saturated due to either or both of two processes: cooling of the air and adding water vapour. With sufficient saturation, precipitation will fall to the surface; an exception is virga, which evaporates before reaching the surface.

Clouds in the troposphere, the atmospheric layer closest to Earth's surface, have Latin names due to the universal adaptation of Luke Howard's nomenclature. It was introduced in December 1802 and became the basis of a modern international system that classifies these tropospheric aerosols into several physical forms, then cross-classifies them as low-, middle- and high-étage according to cloud-base altitude range above Earth's surface. Clouds with significant *vertical* extent occupying more than one étage are often considered a distinct group or sub-group. One physical form shows free-convective upward growth into low or vertical *heaps* of cumulus (*cumuliform*).

Other forms appear as non-convective layered *sheets* like low stratus (*stratiform*), and as limited-convective *rolls* or *ripples* as with stratocumulus (*stratocumuliform*). Both of these layered forms have middle- and high-étage variants identified respectively by the prefixes *alto-* and *cirro-*. Thin fibrous *wisps* of cirrus are a physical form found only at high altitudes of the troposphere (*cirriform*).

In the case of clouds with vertical extent, prefixes are used whenever necessary to express variations or complexities in their physical structures. These include *cumulo-* for complex highly convective vertical *nimbus* storm clouds (*cumulonimbiform*), and *nimbo-* for thick stratiform layers with sufficient vertical depth to produce moderate to heavy *precipitation*. This process of cross-classification produces ten basic genus-types or genera, most of which can be subdivided into *species* and *varieties*. Synoptic surface weather observations use code numbers to record and report any type of tropospheric cloud visible at scheduled observation times based on its height and physical appearance.

While a majority of clouds form in Earth's troposphere, there are occasions when they can be observed at much higher altitudes in the stratosphere and mesosphere. Clouds that form above the troposphere have common names for their main types, but are sub-classified *alpha-numerically* rather than with the elaborate system of Latin names given to cloud types in the troposphere. These three main atmospheric layers that can produce clouds, along with the lowest part of the cloudless thermosphere, are collectively known as the homosphere. Above this lies the heterosphere (which includes the rest of the thermosphere and the exosphere) that marks the transition to outer space. Clouds have been observed on other planets and moons within the Solar System, but, due to their different temperature characteristics, they are often composed of other substances such as methane, ammonia, and sulphuric acid as well as water.

Etymology

The origin of the term *cloud* can be found in the old English *clud* or *clod*, meaning a hill or a mass of rock. Around the beginning of the 13th.century, it was extended metaphorically to include rainclouds as masses of evaporated water in the sky because of the similarity in appearance between a mass of rock and a cumulus heap cloud. Over time, the metaphoric term replaced the original old english*weolcan* to refer to clouds in general.

Bibliography

Ahmad, Y. Hassan and Donald Routledge Hill: *Islamic Technology: An illustrated history*, Cambridge University Press, 1986.

Arthur, Getis; Judith Getis and Jerome D. Fellmann: *Introduction to Geography*, McGraw-Hill, New York City, United States, 2000.

Baumann, Edward: *Applications of Neon Lamps and Gas Discharge Tubes*, Carlton Press, New York, 1966.

Bluestein, H.: *Synoptic-Dynamic Meteorology in Midlatitudes: Principles of Kinematics and Dynamics*, Oxford University Press, 1992.

Brasseur, Guy P.; Orlando, John J.; Tyndall, Geoffrey S.: *Atmospheric Chemistry and Global Change*, Oxford University Press, 1999.

Colin, Martin; Geoffrey, Parker: *The Spanish Armada*, Manchester University Press, London, 1999.

Dart, RC.: *Medical toxicology*, Williams & Wilkins, Philadelphia, 2004.

Derek, Piggott: *Gliding: a Handbook on Soaring Flight*, Knauff & Grove, Columbia, 1997.

Donald, C. Ahrens: *Meteorology Today,* Brooks/Cole Publishing, United States, 2006.

Erik A. Rasmussen, John Turner: Polar Lows, Cambridge University Press, London, 2003.

Finlayson-Pitts, Barbara J.; Pitts, James N., Jr.: *Chemistry of the Upper and Lower Atmosphere*, Academic Press, London, 2000.

Ford MD, Delaney KA, Ling LJ, Erickson T.: *Clinical toxicology*, WB Saunders Company, USA, 2001.

Garratt, J.R.: *The atmospheric boundary layer*, Cambridge University Press, United Kingdom, 1992.

Graham, Howard Johnson, Martin: *High-speed signal propagation : advanced black magic,* Prentice Hall PTR, Upper Saddle River, NJ, 2002.

Greenwood, Norman N.; Earnshaw, Alan: *Chemistry of the Elements,* Butterworth-Heinemann, United Kingdom,1997.

Holton, J.R.: *An Introduction to Dynamic Meteorology,* Elsevier, Burlington, Md, 2004.

Housecroft, C. E.; Sharpe, A. G.: *Inorganic Chemistry,* Prentice Hall, United States, 2004.

J. Schopf: *Earth's Earliest Biosphere: Its Origin and Evolution,* Princeton University Press, Princeton, N.J., 1983

Jacobson, Mark Z.: *Fundamentals of Atmospheric Modeling,* Cambridge University Press, New York, 2005.

K.E.K. Neuendorf; J.P. Mehl, Jr., and J.A. Jackson: *Glossary of Geology,* Springer-Verlag, New York, 2005.

Linda T. Elkins-Tanton: *Uranus, Neptune, Pluto, and the Outer Solar System,* Chelsea House, New York, 2006.

Lutgens, Frederick K. and Edward J. Tarbuck: *The Atmosphere,* Prentice Hall, United States 1995.

Mark Zachary, Jacobson: *Fundamentals of Atmospheric Modeling,* Cambridge University Press, 2005.

Michael A. Mares: *Encyclopedia of Deserts,* University of Oklahoma Press, United States, 1999.

Owen, Lewis A.; Pickering, Kevin T.: *An Introduction to Global Environmental Issues*: Taylor & Francis, New Delhi, 1997.

Paul, Clayton R.: *Fundamentals of Electric Circuit Analysis,* John Wiley & Sons, United States, 2001.

Peebles, Peyton, *Radar Principles,* John Wiley & Sons, New York, 1998.

R. R. Rogers and M. K. Yau: *Short Course in Cloud Physics,* Butterworth-Heinemann, United Kingdom, 1989.

Robert, Harrison: *Large Wind Turbines,* John Wiley & Sons, Chichester, 2001.

Salomons, Erik M.: *Computational Atmospheric Acoustics,* Kluwer Academic Publishers, Netherlands, 2001.

Serway, Raymond A.; Jewett, John W.: *Physics for Scientists and Engineers,* Brooks/Cole Publishing, United States, 2004.

Sharad K. Jain, Pushpendra K. Agarwal, Vijay P. Singh: *Hydrology and Water Resources of India,* Springer, United Kingdom, 2007.

Tchobanoglous, George: *Wastewater Engineering,* McGraw Hill, New York, 1979.

Todd S. Glickman: *Glossary of Meteorology,* American Meteorological Society, Boston, 2000.

Walter, J. Saucier: *Principles of Meteorological Analysis,* Courier Dover Publications, US., 2003.

Wayne, Richard P.: *Chemistry of Atmospheres,* Oxford University Press, London, 2000.

Index

❑❑❑